Lecture Notes in Economics and Mathematical Systems

380

Jan Odelstad

Invariance and Structural Dependence

Springer-Verlag

Berlin Heidelberg New York
London Paris Tokyo
Hong Kong Barcelona
Budapest

Author

Dr. Jan Odelstad
Department of Philosophy
Uppsala University
Villavägen 5
S-752 36 Uppsala, Sweden

This is a revised version of a doctoral thesis, submitted in mimeographed form to the
Faculty of Arts, Uppsala University, 1988.

ISBN 978-3-540-55260-4   ISBN 978-3-642-48388-2 (eBook)
DOI 10.1007/978-3-642-48388-2

Typesetting: Camera ready by author

42/3140-543210 - Printed on acid-free paper

PREFACE

This is a revised version of a doctoral thesis, submitted in mimeographed form to the Faculty of Arts, Uppsala University, 1988. It deals with the notions of structural dependence and independence, which are used in many applications of mathematics to science. For instance, a physical law states that one physical aspect is structurally dependent on one or more other aspects. Structural dependence is closely related to the mathematical idea of functional dependence. However, structural dependence is primarily thought of as a relation holding between aspects rather than between their measures.

In this book, the traditional way of treating aspects within measurement theory is modified. An aspect is not viewed as a set-theoretical structure but as a function which has sets as arguments and set-theoretical structures as values. This way of regarding aspects is illustrated with an application to social choice and group decision theory.

Structural dependence is connected with the idea of concomitant variations and the mathematical notion of invariance. This implies that the study of this notion has roots going back to Mill's inductive logic, to Klein's Erlangen Program for geometry and to Padoa's method for proving the independence of symbols in formal logic.

A distinction between dependence in the sense of determination and in the sense of relevance is drawn. To study determination the idea of preservation of equality is applied to two different "levels": the object level, which gives rise to the notion of conformity, and the structure level, which gives rise to a notion of subordination. The study of relevance is primarily based on the study of determination.

Structural dependence both in the sense of determination and of relevance can be found in degrees from complete determination and relevance to complete undetermination and irrelevance. The possibility of different "scales" of determination and relevance is considered.

The structural dependence relations which are found in physics seem to be of a special kind. In other sciences, the dependence relations may be of a quite different character. One chapter is devoted to a study of the dependence relation between the social ordering and the individual orderings in social choice and group decision theory.

I am grateful to many people for help with my dissertation. Stig Kanger was my supervisor until his untimely death in March 1988. He aroused my interest in measurement and decision theory. His two essays on measurement have had a substantial influence on me, as have our many discussions on these and other philosophical topics. He encouraged my work on structural dependence and we were planning its continuation at the time of his death. I regret that I was not able to discuss the final version with him.

After Stig Kanger's death, Sören Stenlund kindly agreed to oversee my work. He has provided valuable suggestions on parts of the manuscript and assisted in many practical matters, and I am grateful for all his encouragement. Sten Lindström read the whole manuscript, some parts in several versions, and made numerous suggestions for further developments and improvements. Lars Lindahl has offered constructive criticism and stimulating discussions on some parts of the manuscript. I am also grateful to Lars for his longstanding support and encouragement. Professor Jan Berg was the Faculty Opponent for the public examination of the dissertation. I am grateful to him for his detailed criticism and valuable suggestions. Various parts of the book have been ventilated in seminars at the Philosophy Department in Uppsala. I am grateful to the participants for constructive criticism and suggestions. In this connection I want especially to mention Lars Bergström, Sven Danielsson, Bengt Molander, Wlodzimierz Rabinowicz and Bertil Strömberg. In writing chapter 3 of this essay I have profited considerably from discussions with Sven Danielsson and Margareta Sjöberg back in the 70's.

Paul Needham has spent much time and effort in bringing order to my unstructured English and arriving at a suitable terminology for the various notions introduced in this essay. Bengt Molander and Sten Lindström have assisted with trouble shooting and in various other ways in connection with producing this document on a computer. During the academic year 1986-87 the work on this book was financially supported by the Swedish Council for Building Research, project number 860478-0. I wish to thank Senior Research Officer Eva Fredell for all her assistance and interest.

Special thanks are due to my wife Lena. Without her interest and encouragement the work on this book would never have reached its present stage. Finally, many thanks to my mother and my mother-in-law for all their baby-sitting during a critical period.

I dedicate this book to the memory of my father.

Uppsala, August 1991

Jan Odelstad

TABLE OF CONTENTS

PART ONE

PROBLEM AREA AND BASIC FORMAL APPARATUS

CHAPTER 1

THE CONCEPT OF DEPENDENCE IN APPLIED

MATHEMATICS; A FIRST ACCOUNT

1.0   Introduction

This essay is about the scientific use of *dependence* and *independence* and other related concepts. There are many notions of dependence and independence used in science but only some of them will be studied here. One important notion of independence which I shall not deal with is the notion of *independent events* in probability theory. The notion of *causal dependence* is another important notion I shall leave out of account. The subject matter is rather the notions of dependence and independence which are central in most applications of mathematics to science.

By way of outlining the landscape through which our investigation will take us, I shall present two quotations, one from a classical mathematics text, and one from a modern textbook of mathematical physics.

> Suppose two quantities which are susceptible of change so connected that if we alter one of them there is a consequent alteration in the other, this second quantity is called a *function* of the first. ... If a function of x is supposed equal to another quantity, as for example sinx=y, then both quantities are called *variables,* one of them being the *independent variable* and the other the *dependent variable.* An *independent variable* is a quantity to which we may suppose any value arbitrarily assigned; a *dependent* variable is a quantity the value of which is determined as soon as that of some independent variable is known. (Todhunter, 1852, p. 1)

> The notion of a relation existing between the values of two variable physical quantities presents itself immediately in the study of any branch of physics, chemistry or mathematics. For example, the pressure p of a gas at a given temperature is related to the density ρ of the gas; the period of oscillation T of a simple pendulum depends on the length l of the pendulum; the time t that a body takes to fall from rest under gravity depends on the height h from which it is dropped. In all these simple examples it is possible to find a definite

mathematical formula relating the two quantities; the three formulae being the well-known relations

(i) $p = k\rho$,   (ii) $T = 2p\sqrt{(l/g)}$,   (iii) $h = 1/2gt^2$.

In these formulae the quantities $p, \rho; T, l; h, t$ are assumed to be able to take different real values; such quantities are said to be *real variables* or more simply *variables*. The other quantities k, g do not change their values and these are referred to as *constants*. (Chisholm & Morris, 1965, p. 1)

These quotations illustrate how the notions of dependence and independence are traditionally used in mathematics and the applications of mathematics, and the reader is certainly familiar with them. Some points I would like to emphasize are the following.

Dependence is a relation between variable quantities (variables). *Dependent* and *independent* are adjectives attributed to variables. The dependent variable depends on the independent variable or variables. If a variable y is a function of another variable x, then y is dependent on x and the dependence relation, i.e. the connection between x and y, is a function.

My intention is not to track down one single concept of dependence and one of independence. I think that this is hardly possible—even if one restricts oneself to the use of dependence and independence in contexts of which the above quotations are examples, it is hardly a matter of only one notion of dependence and one of independence. Rather I will try to make some of the ideas about dependence/ independence clearer by introducing a number of technically well-defined concepts that will formulate different aspects of the cluster of ideas that is associated with the notions of dependence and independence. The object of this study is thus to construct notions which can be used as tools in analysis and not to conduct an analysis of some special concept of dependence.

I emphasize the introductory character of this chapter. It consists of a preliminary outline of some of the main themes we will be dealing with. I have tried to be as informal as possible and to this end I have allowed myself to make a lot of simplifications and to express myself in general terms. Still, I have on occasion found no alternative but to use a minimum of the formal apparatus which is not properly introduced until chapter 2. The reader will hopefully be able to follow the train of thought on the basis of his previous knowledge and general intuition. The remaining chapters of this essay contain explications, developments and discussions of what is said in chapter 1.

1.1 Determination and relevance

As a first step into the domain of ideas of dependence we look at a simple example from physics. Let me quote a passage from my secondary school physics text.

> The vessels A,B,C and D shown in [the figure below] have different shapes but the same area at the base, and they are filled to the same height with water. The pressure at the bottom is the same in all the vessels since the water pressure depends only on its depth. And since the base area is the same, the force on the bottom (the product of pressure and area) is the same in all four cases. The water in vessels B and C therefore gives rise to a force which is greater than the water's own weight. This result would seem to be unreasonable (paradoxical) and is therefore called *the hydrostatic paradox*.

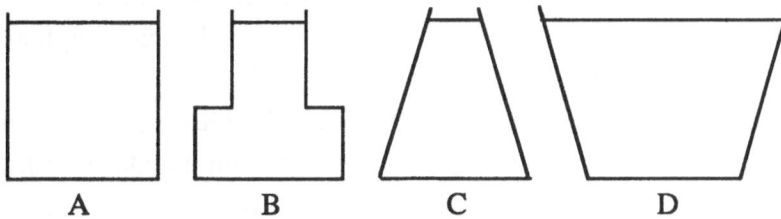

(Bergholm, 1963, p. 40.)

I won't remind the reader of the explanation of the "paradox"—it is of no importance here. The "paradox" will instead be used to illustrate one standard use of the notions of dependence and independence.

We are interested in the force on the bottom of a vessel filled with water. Preliminary considerations may suggest that the following factors may be of importance for determining the force on the bottom :

* the area of the bottom
* the amount of water
* the depth (the distance between the surface and the bottom)
* the shape of the vessel

The factors listed above are to be called *aspects*. They are aspects of the same kind of objects, namely liquid filled vessels. The four vessels in the figure above have equal bottom areas and the distances between the surface and the bottom are equal (i.e. the depths are equal). But the shape and the amount of water differ from vessel to vessel. Interestingly enough, the forces on the bottom are equal in the four vessels. This means that the force on the bottom is dependent only on the depth and the bottom area and thus independent of the amount of water and the shape of the vessel. However, note that we have assumed that all the vessels in the figure above are filled with water, so the density of the liquid is held constant. That is why it is said in the quotation that the liquid pressure depends only on the depth. Generally, the liquid pressure equals density (specific weight) times depth. Hence, if we do not consider cases where the density of the liquid is held constant, then the force on the bottom is dependent on the density of the liquid too. In a more general case, the force on the bottom is dependent on the density, the depth and the bottom area but is still independent of the amount of liquid in the vessel and the shape of the vessel. The force is thus dependent only on density, depth and bottom area. (It is important to note here that even in this more general case many factors are presupposed to be held constant, for example the distance to the center of the earth and the atmospheric pressure. It is common practice in physics to let such implicit presupposition be clear from the context.)

I think it is possible to distinguish between two ideas connected with the concept of dependence, namely the idea of determination and the idea of relevance. They can be introduced in a very preliminary way as follows. The force on the bottom is *determined* (completely) by density, depth and bottom area since only these factors are relevant. The depth is *relevant* to the force because the force is determined by density, depth and bottom area but not by just density and bottom area. Hence, the depth is an essential part of what the force is determined by. On the other hand, the amount of water and the shape of the vessel are not relevant to the force. Determination and relevance are thus related. α is relevant to β when β is determined by α and possibly some other aspects but not by these other aspects alone. In other words, α is relevant to β if β is determined by α or there are $\alpha_1,...,\alpha_n$ such that β is determined by $\alpha,\alpha_1,...,\alpha_n$ but β is not determined by $\alpha_1,...,\alpha_n$. And β is determined by $\alpha_1,...,\alpha_n$ when $\alpha_1,...,\alpha_n$ are the only factors which are relevant to β. Determination thus seems to be closely connected with the notion of being *dependent only upon*: β is dependent only upon $\alpha_1,..,\alpha_n$ iff β is determined by $\alpha_1,..,\alpha_n$. I think that is not uncommon to omit "only" and with "β is dependent on $\alpha_1,...,\alpha_n$" mean "β is determined by $\alpha_1,...,\alpha_n$". It is of course important to distinguish this from "β is dependent on α" in the sense " α is relevant to β".

There is another idea of relevance which could be illuminating. According to this, the depth is relevant to the force because if "everything else" besides force and depth is kept constant, then the force is not constant and is determined by the depth. More specifically, if density and bottom area are kept constant then the force is not necessarily constant and is determined by the depth. Relevance is then connected to determination in the following way. α is relevant to β when β is not constant and is determined by α, when everything else is kept constant. The condition that β is not constant is important. It implies that to some extent β varies when α varies. There are therefore cases where α has a definite influence on β. It is thus a prerequisite for α being relevant for β that α "affects" β. From this it follows that the amount of water is not relevant to the force because if everything else is constant, i.e. if density, depth and bottom area are constant, then the force is constant even if the amount of water varies. Of course, it is a central question here just what exactly is meant by variation and constance. We shall return to this subject in later chapters.

If β is determined by $\alpha_1,...,\alpha_n$ then it is common to say that β is a function of $\alpha_1,...,\alpha_n$. Hence, the force on the bottom is a function of the density, the depth and the bottom area. Note that β is a function of $\alpha_1,...,\alpha_n$ if and only if β is dependent only upon $\alpha_1,...,\alpha_n$. But note that even if β is a function of $\alpha_1,...,\alpha_n$ it is not necessary that α_1 is relevant for β. For β can be a function of only $\alpha_2,...,\alpha_n$ and, hence, α_1 is completely irrelevant for β. From the fact that β is dependent only upon $\alpha_1,...,\alpha_n$ it does not follow that β is dependent on each and every one of $\alpha_1,..,\alpha_n$.

This essay is to a large extent a study of determination and relevance. We shall especially be concerned with the fact that both determination and relevance seem to come in degrees. β can be completely determined by α or just partially determined. And α can be strongly relevant for β or just weakly relevant. We shall try to characterize determination and relevance of different strength—and of course try to define undetermination and irrelevance.

In attempting to explicate determination use is made of a single idea applied to two different levels of abstraction. This idea amounts to *preservation of equality*. Applied to *objects* this idea will give rise to one of our central notions, *conformity*. Applied to *structures* it will give rise to another central notion, *subordination*. Let us first attempt a preliminary sketch of what preservation of equality means on the more concrete level of abstraction, i.e. for objects.

The force on the bottom of a vessel filled with a liquid is determined by the density of the liquid, the depth (the distance between the surface and the bottom) and the bottom area. This means that if we have two vessels filled with liquids of the same density, with the same depth and with the same bottom area, then the forces on the bottoms are the same for both vessels. Hence, equality in respect of liquid density, depth and bottom area implies equality in respect of force on the bottom. More generally, an aspect β is determined by the aspects $\alpha_1,...,\alpha_n$ if and only if whenever the objects a and b are equal with respect to $\alpha_i, 1 \leq i \leq n$, then a and b are equal with respect to β. It is obvious here in what sense determination involves equality preservation; equality with respect to every $\alpha_i, 1 \leq i \leq n$, is preserved in equality with respect to β.

As was said before we often use "is a function of" instead of "determined by". Thus, "is a function of" is a relation between aspects. But we also use "is a function of" as a relation between measures. Instead of saying that the force is a function of the density, the depth and the bottom area we can say that the force in Newtons, φ, is a function of the density in kg/m³, δ, the depth in m (meters), η, and the bottom area in m², μ. φ,δ,η and μ are here measures (scales) and as such real-valued functions with the class K of liquid filled vessels as domains. That φ is a function of δ,η and μ is usually taken to mean that there is a function $F:Re^3 \rightarrow Re$ such that for all x in K

$$\varphi(x) = F(\delta(x), \eta(x), \mu(x)).$$

One often comes across the opinion that "is a function of" is only a relation between measures and should be understood in the way just stated. The use of the expression in the context of aspects is then simply regarded as elliptical. But as we shall see in chapter 6, functional dependence between measures is (under general conditions) equivalent to functional dependence between aspects (in the sense of equality preservation).

We are now in a position to look at the meaning of complete undetermination. Again we begin with a first preliminary outline, in conformity with the above sketch of complete determination.

The force on the bottom is completely undetermined by the density alone, because given the density of a liquid we cannot say anything at all about the force as long as we have said nothing about the depth and the bottom area. Analogously, the force on the bottom is completely undetermined by each of depth and bottom area taken alone. (We presuppose here that we are studying liquid filled vessels, so the measures of density, depth and bottom area are always distinct from zero.) Generally, an aspect β is completely undetermined by an aspect α if

(1) whatever holds of an object with respect to α (i.e. how much the object has of α)

does not imply anything at all about

(2) whatever holds of the object with respect to β (i.e. how much the object has of β).

Alternatively, we can explain this by saying that β is completely undetermined by α if for all objects x and y of which α and β are aspects, it is possible that there is an object z such that z is equal to x in respect of α and z is equal to y in respect of β.

We can also use measures to explain complete undetermination. Let m_α be a measure of α and m_β a measure of β. β is undetermined by α if the value of m_β for an object x, i.e. $m_\beta(x)$, is not restricted by the value of m_α for x, i.e. $m_\alpha(x)$. Note that if β is completely determined by α then $m_\beta(x)$ is determined by $m_\alpha(x)$; given $m_\alpha(x)$ there is only one possible value for $m_\beta(x)$. If φ, δ, η and μ are measures as described above then $\varphi(x)$ is determined when $\delta(x)$, $\eta(x)$ and $\mu(x)$ are given, which is what we expect since φ is a function of δ, η and μ. But $\varphi(x)$ is not restricted when $\delta(x)$ is given as far as $\eta(x)$ and $\mu(x)$ are not given.

The explication of determination in terms of equality preservation for objects is one way of interpreting the idea of determination and will be investigated in detail in chapter 6. However, as has been pointed out already, there is another way the idea of equality preservation might be applied. According to this, equality preservation is applied to structures and not to objects. We can use equality preservation for structures as the basis for an explication of determination which will be quite different from the explication in terms of equality preservation for objects. While equality preservation for objects give rise to a rather weak notion of determination, the one founded on equality preservation for structures is quite strong. However, the relations between the two notions are somewhat involved. A detailed study of determination as equality preservation for structures is one of our main tasks in this essay. In this introductory chapter we can only give a very rough sketch of what this notion amounts to.

Let α be an aspect of all the elements in the set X and let A be a subset of X. Then α induces a structure on A. It may for example hold that x is greater than y with respect to α and that z is twice as large as u with respect to α. The structure which α induces on A we call the α-structure on A. Let us now suppose that β is also an aspect of all the elements in the set X. The idea is now that β is determined by α if equality between the α-structures on A and on another subset B of X implies equality between the β-structures on A and on B. We can say that the α-structure on A states what holds for the objects in A with respect to α. With this inter- pretation β is determined by α if the same thing's holding of A and B with respect to α implies that the same thing's holding of A and B with respect to β. To make this idea precise we shall use the algebraic notion of an isomorphism and the concept of determination we then arrive at will be called *subordination*. An aspect β is subordinated to an aspect α if the following holds: whenever φ is an isomorphism between the α-structure on A and B then φ is an isomorphism between the β-structure on A and B.

To make this idea somewhat more concrete we return to the example at the beginning of this section. Let A be a set of vessels filled with water. For every vessel x in A, let x_2 be a vessel filled with water such that its depth is double that of x but with the same bottom area. Let A_2 be the set of all such x_2s. Suppose φ is the correspondence between A and A_2 such that $\varphi(x)=x_2$. It is obvious that φ is an isomorphism between the bottom area structure of A and of A_2 since x and $\varphi(x)$ have equal bottom areas. φ is also an isomorphism between the depth structure of A and A_2 since (1) depth is measured on a ratio scale and (2) if the depth of x is k times the depth of y then the depth of x_2 is k times the depth of y_2. (2) holds because the depth of x_2 is double the depth of x and analogously for y_2 and y. φ is also an isomorphism of the force-on-the-bottom structure of A and of A_2. This holds for the following reason. The force on the bottom is depth times bottom area, so the force on the bottom of $\varphi(x)$ is twice the force on the bottom of x. Hence, if the force in x is k times the force in y then the force in $\varphi(x)$ is k times the force in $\varphi(y)$. And, since force is measured on a ratio scale, φ is an isomorphism between the force structure of A and A_2.

We have given some idea here of how we can prove that φ, which is an isomorphism between the density structure of A and of A_2 and at the same time an isomorphism between the bottom area structure of A and of A_2, is also an isomorphism between the force structures of A and of A_2. It holds generally that if ψ is an isomorphism between the density structure of X and Y and at the same time an isomorphism of the bottom area structure of X and Y, then ψ is an isomorphism between the force structure on X and Y. The force on the bottom is therefore subordinate to depth and bottom area—N.B. we are considering vessels filled with liquid of the same density.

A more detailed study of dependence between aspects demands of course a precise view of what an aspect is. It is common in modern measurement theory to regard an aspect as a relational system and I do so in this study, too. But the concept of a relation I shall use in connection with aspects is not the usual one. In chapter 3 the alternative relation concept will be presented in detail. A preliminary sketch of it is given in section 8 of this chapter.

1.2 Partial determination

In the preceding section I maintained that dependence, both in the form of determination and relevance, comes in degrees, and a very preliminary sketch of the meaning of complete determination was presented. In this section some examples of partial determination are considered.

The archbishop of Sweden is elected according to the following rule. Members of the priesthood of the archdiocese and of a committee of laymen each vote for the candidate they prefer, and the candidates are ordered according to the number of votes they get, after which the government chooses one of the first three candidates.

Let R be the relation that rank orders the candidates according to the number of votes they get. And let C be the one-place relation "chosen by the government" defined on the set of candidates. Since C consists of one of the three candidates highest ranked by R, C is partly but not completely determined by R. If we know the rank order we know something of the result of the choice but cannot—in the general case—specify it completely. And, conversely, if we know which candidate the government has chosen we have partial knowledge of the rank order since we know that this candidate must be one of three highest ranked.

In chapter 5 we shall illustrate some of the central concepts of this study with group decision theory. The group preference relation is usually considered to be dependent on the individual preference relations. But should the group preference relation be completely determined by the individual preference relations or just partly determined? Or are the individual preference relations only relevant for the group preference? The answer is not as obvious as it may seem and we shall discuss it further in chapter 5.

Among elementary physical laws partial determination does not seem to be common. In the preceding section we saw that the force on the bottom of a liquid filled vessel is dependent on the density, the depth and the bottom area, but independent of the amount of water in the vessel and the shape of the vessel. We can express this in terms of determination and relevance in the following way: The force on the bottom is completely determined by density, depth and bottom area while the amount of water and the shape of the vessel are completely irrelevant.

Note however that the force on the bottom is completely undetermined by each of the density, depth and bottom area taken alone. Given the density we cannot say anything at all about the force as long as we have said nothing about the depth and the bottom area. (We presuppose here that we are studying liquid filled vessels, so the measures of density, depth and bottom area are always distinct from zero.) Density, depth and bottom area are, however, relevant for the force on the bottom, which has already been pointed out.

In this example the dependent variable, i.e. the force on the bottom, is completely determined by the independent variables—i.e. the density, depth and bottom area—taken together, but completely undetermined by each one of the independent variable taken alone. This is not an uncommon situation in physics. Let us now, however, look at a fictitious example where this is not the case.

Suppose that m_i is a measure of α_i for all i, $0 \leq i \leq 2$, and that the α_i are aspects of the same class of objects X. Suppose further that for all x in X

$$m_0(x) = f(m_1(x), m_2(x)),$$

where f is a function from Re×Re to Re such that

$$f(r,s)=0 \quad \text{if } r \leq 0$$
$$f(r,s)=s \quad \text{if } r > 0.$$

We might suppose that m_0 is the concentration in kg/dm^3 of one substance S, m_2 the concentration in kg/dm^3 of another substance T and m_1 the temperature in °C (degrees Celsius), when a chemical reaction has reached equilibrium. The functional connection between m_0, m_1, and m_2 given above implies that when the temperature is below the freezing-point the substance S does not occur.

It is obvious that $m_0(x)$ is determined when $m_1(x)$ and $m_2(x)$ are given. This implies that, with the explication of complete determination in terms of equality outlined in the preceding paragraph, α_0 is completely determined by α_1 and α_2. But α_0 is not completely undetermined by α_1 taken alone, because if $m_1(x) \leq 0$ then $m_0(x)=0$ independent of the value of $m_2(x)$. Thus, there is no object y in X such that $m_0(y)>0$ and $m_1(y) \leq 0$.

1.3 Structural dependence

We have tentatively proposed the following as one meaning of the phrase "the force on the bottom is determined by the density of the liquid, the depth and the bottom area": If we have two vessels with the same bottom area, filled with liquid of the same density to the same depth, then the forces on the bottoms are the same for both vessels. Now note that the following also holds: If we have two vessels with the same bottom area and filled with liquids of the same density, and they have the same force on the bottom, then the depths are the same for the two vessels. The depth is thus determined by the density, the bottom area and the force on the bottom. If the density and the bottom area is kept constant then the depth is determined by the force on the bottom. Therefore, if we accept the relation between determination and relevance sketched above, then the force on the bottom is relevant for the depth. The depth is thus dependent on the force on the bottom when dependence is taken in the sense of relevance.

This may seem strange. We usually think that we can manipulate (vary) depth and—given that the density and the bottom area are kept constant—thereby change the force on the bottom, but not manipulate (vary) the force on the bottom and thereby change the depth. But it is important at this point to note that the notion of dependence—in the sense of determination—which will be investigated here is a symmetric relation: if α is dependent on β then β is dependent on α. However, *wholly dependent* is not a symmetric relation: it may be the case that α is wholly dependent on β but that β is not wholly dependent on α (only dependent on α). We shall take a closer look at this in later chapters. The essential thing just now is that the notion of dependence we shall be dealing with is not (directly) connected with manipulability.

In applied mathematics this symmetric dependence concept is not uncommon. As we have pointed out above,
(1) the force on the bottom is a function of the density, the depth and the bottom area.
But it is also quite natural to say that
(2) the depth is a function of the density, the bottom area and the force on the bottom.
Let us, in preliminary fashion, say that a variable quantity y is a function of other variable quantities $x_1,...,x_n$ if a rule is laid down which determines one value of y when the values of $x_1,...,x_n$ are given. Suppose y is a function of x_1 and x_2. Then it may be the case that given the rule which determines one value of y for given values of x_1 and x_2, we can find a rule which determines one value of x_1 when one value of y and x_2 are given. When this happens x_1 is a function of y and x_2 and it seems adequate to say that x_1 is dependent on y and x_2 irrespective of whether we can

manipulate (vary) y.

The notion of dependence in the sense of determination we are interested in is thus in a sense symmetric. However, this does not seem to be the case for the notion of relevance we shall study and which can be explicated in terms of determination. But it is not asymmetric either. The symmetry of partial determination gives relevance an "undirected" character, to use a rather vague expression, and this distinguishes it from other notions of relevance. The notion of dependence of interest here can be contrasted with that of *causal dependence*. There are lots of different views of causality and I will in this context avoid taking a standpoint of my own to this classical problem. But I think that the following statements are in accordance with at least some views of causality. Both in the sense of determination and relevance causal dependence is essentially "directed". It is the cause that determines the effect, and the effect doesn't even partially determine the cause. And if X is causally relevant to Y the converse does not normally hold.

In this essay little will be said about *dependent* and *independent* as attributes of variables. Instead the focus is on the relation *dependent on* between variable quantities. But a few words about the attributes and the connection between them and the relation are in order.

In (1) above the force on the bottom is the dependent variable while the density, the depth and the bottom area are the independent variables. However, in (2) the depth is the dependent variable while the density, the bottom area and the force on the bottom are the independent variables. Todhunter gives the following characterization of dependent and independent variable:

> An *independent variable* is a quantity to which we may suppose any value arbitrarily assigned; a *dependent* variable is a quantity the value of which is determined as soon as that of some independent variable is known. (Todhunter, 1852, p. 1)

And then Todhunter makes the following important observation:

> Frequently when we are considering two or more variables it is in our power to fix upon whichever we please as the *independent* variable, but having once made our choice we must admit no change in this respect throughout our operations; at least such a change would require certain precautions and transformations. (Todhunter, 1852, p. 1)

I think that the classification of variables as dependent or independent is normally relativized to the actual situation; a variable can be dependent in one situation and independent in another. The situations are often of an experimental character. Jevons calls the dependent variable a "variant" and the independent variable just a "variable", and discusses their relative character as follows:

> Almost every series of quantitative experiments is directed to obtain the relation between the different values of one quantity which is varied at will, and another quantity which is caused thereby to vary. We may conveniently distinguish these as respectively the *variable* and the *variant*. When we are examining the effect of heat in expanding bodies, heat, or one of its dimensions, temperature, is the variable, length the variant. If we compress a body to observe how much it is thereby heated, pressure, or it may be the dimensions of the body, forms the variable, heat the variant. In the thermo-electric pile we make heat the variable and measure electricity as the variant. That one of the two measured quantities which is an antecedent condition of the other will be the variable. (Jevons, 1874, 1924, p. 440)

Given a description of a situation it is usually clear which are the independent and dependent variables, respectively. But in describing the situation in different ways it is often possible to let either of the variables be an independent variable. However, we seldom have the freedom to choose an arbitrary set of the variables as the set of independent variables.

In Krantz et.al. (1971) there is a discussion of the relation between independent and dependent variables in additive conjoint measurement (see p. 276f.). The authors argue for the thesis that "additive conjoint measurement is actually symmetric in the independent and dependent variables". The discussion ends with the following conclusion:

> The choice of independent variables is a convention, from the standpoint of the theory, although there are often good reasons for such conventions, based on other considerations, among them extensive measurement of variables, natural product structures, experimental control of certain variables, or assumptions about error of the type made in linear statistical models. (Krantz et.al., 1971, p. 278)

It may seem that the dependent variable is obviously dependent on the independent one (for simplicity we suppose for the moment that there is only one independent variable), and not the other way round. In one sense of dependence that is certainly so. In this sense dependence is "directed", it has a "direction" from the independent toward the dependent variable. Since, as we have seen above, the characterization of a variable as dependent or independent seems to depend on the situation, the "direction" of the dependence relation likewise depends on the situation. But as we saw at the beginning of this section, there is a use of the dependence relation between variables which is not relativized in this way to a situation, and used in this sense the dependence relation lacks "direction". That partial determination has the property of symmetry emphasizes this.

To see dependence as "undirected" implies of course that important aspects of what we often mean by "dependent" are not treated. The notion of dependence under study is therefore a meagre notion, but as far as I can see not without interest since it is used in the application of mathematics. Let us call this notion *structural dependence,* because it focuses on the structural relation between aspects (variables). We shall look in more detail at the reasons motivating this choice of term later on.

1.4 Dependence and concomitant variations

The notions of dependence of interest here are intimately associated with the idea of concomitant variations. It is natural at this point to return to John Stuart Mill.

In *System of Logic* from 1843 Mill presents five methods of experimental inquiry by which we discover and demonstrate causal relationships. The first four methods are used in studying the effect of eliminating or isolating factors. As Mill points out, these methods do not work when we come across

> Permanent Causes, or indestructible natural agents, which it is impossible either to exclude or to isolate; which we can neither hinder from being present, nor contrive that they shall be present alone. (Mill, 1843, p. 431)

But, as Mill declares,

> we have still a resource. Though we cannot exclude an antecedent altogether, we may be able to produce, or nature may produce for us, some modification of it. By a modification is here meant, a change in it, not amounting to its total removal. (Mill, 1843, p. 433.)

To clarify this line of thought Mill appeals, among other things, to the following example.

> Let us now suppose the question to be, what influence the moon exerts on the surface of the earth. We cannot try an experiment in the absence of the moon, so as to observe what terrestrial phenomena her annihilation would put an end to; but when we find that all the variations in the *position* of the moon are followed by corresponding variations in the time and place of high water, the place being always either the part of the earth which is nearest to, or that which is most remote from, the moon, we have ample evidence that the moon is, wholly or partially, the cause which determines the tides. (Mill, 1843, p. 434.)

The method used in this example Mill calls "the Method of Concomitant Variations" and it is regulated by the following canon.

> Whatever phenomenon varies in any manner whenever another phenomenon varies in some particular manner, is either a cause or an effect of that phenomenon, or is connected with it through some fact of causation. (Mill, 1843, p. 435)

The method of concomitant variations is not only useful in cases in which the other methods are not applicable but also as a complement to them.

> When by the Method of Difference it has first been ascertained that a certain object produces a certain effect, the Method of Concomitant Variations may be usefully called in to determine according to what law the quantity or the different relations of the effect follow those of the cause. (Mill, 1843, p. 437)

According to Mill inductive investigations are a search for causes, which explains why he emphasized that the five methods of experimental inquiry are methods for discovering and demonstrating causal relationships. However, it has often been said that the search for causal relationships does not play such a predominant role in science as Mill thought. The method of concomitant variations can of course be used for determining relationships between phenomena which may not be causal. W.E. Johnson discusses the methods in connection with a conception of dependence which he applies to variations of phenomenal characters (see Johnson, 1922, p. 219f.) and Susan Stebbing in connection with concepts such as functional correlation and functional dependence (see Stebbing, 1933, p. 352). J.O. Wisdom formulates a criterion of functional dependence which he sees as one form of the criterion of co-variation originally given by Mill as the method of concomitant variations. The criterion of functional dependence has the following wording:

> *Criterion of Functional Dependence:* In a given situation, if variation of X is accompanied by some variation of E, such that the series of corresponding values (X,E) fit a smooth curve within the limits of experimental error, then it is probable that E is a simple function of X, *i.e.* E=f(X).
> This may be illustrated by Charles's law that changes in the volume of a fixed mass of gas at constant pressure are proportional to changes in temperature. ...
> X and E may be parallel effects of an anterior cause, or not, but in either event, X is necessary and sufficient to determine E (in the mathematical sense of determine according to which when the value of X is known that of E can be calculated), given the conditions in which the corresponding values of X and E are found. (Wisdom, 1945, pp. 335-336)

In the revised form given by Johnson, Stebbing, Wisdom and others, Mill's method of concomitant variations is applicable to the notions of dependence we have discussed in very preliminary fashion in the preceding sections. I think this is particularly clear from the presentation of two methods of variation given in Marc-

Wogau (1950) (see p. 190f.). The purpose of the methods of variation is, according to Marc-Wogau, to determine dependence-relations between factors. In the following passage *a,b,c* and *d* denote observed factors while *x* is a summary of existing but non-observed (i.e. unknown) factors. The two methods of variation are the following:

> 1. One starts with a situation *abcdx* and lets the factor *a* vary in strength or magnitude (take on the values $a_1,a_2,a_3,...$) whilst the other factors *b* and *c* are held constant. If it is found that *d* remains constant, i.e. that the following series is obtained:
>
> $$a_1bcdx$$
> $$a_2bcdx$$
> $$a_3bcdx$$
> $$a_4bcdx$$
> $$........,$$
>
> it can be concluded that *d* is probably not dependent on *a*.
> 2. Again, starting from *abcdx* and letting factor *a* vary (take on different values $a_1,a_2,a_3,...$), *b* and *c* are held constant. If it is found that *d* varies (takes on different values $d_1,d_2,d_3,...$), i.e. that the following series is obtained:
>
> $$a_1bcd_1x$$
> $$a_2bcd_2x$$
> $$a_3bcd_3x$$
> $$........,$$
>
> it can be concluded that *d* is probably dependent on *a*. (Marc-Wogau, 1950, pp. 190-191)

Marc-Wogau exemplifies the first method with "the hydrostatic paradox", and I shall follow his presentation. The "hydrostatic paradox" says, as we have already seen, that the force on the bottom of a vessel filled with water does not depend on the amount of water but on the bottom area and the depth. To examine this proposition experimentally we can proceed in the following way. Initially we distinguish four factors or aspects of the water filled vessel: *a* the amount of water, *b* the bottom area, *c* the depth and *d* the force on the bottom. We choose vessels of different shape but with the same bottom area and with the same depth. The amount of water therefore varies, i.e. *a* takes different values $a_1,a_2,a_3,...$. The bottom area (*b*) and the depth (*c*) on the other hand are kept constant. The force on the bottom (*d*) is found to be the same in all cases. Thus we get an instance of the table in part 1 in the quotation from Marc-Wogau above. We conclude that the force on the bottom (*d*) is independent of the amount of water (*a*), i.e. the amount of water is irrelevant for the force on the bottom.

We can also use "the hydrostatic paradox" as an example of the second method. Let *a,b,c,d* be as above. We can vary *c* while keeping *a* and *b* constant if we choose vessels of different shapes such that for all vessels the bottom area and the amount of water are equal but the depth varies.

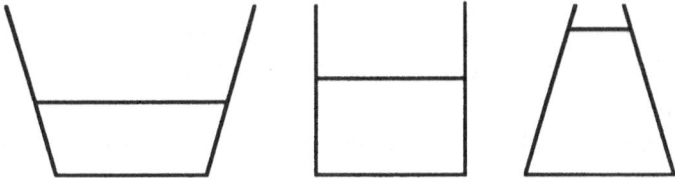

Then we find that the force on the bottom varies too, and obtain the following series:

$$abc_1d_1x$$
$$abc_2d_2x$$
$$abc_3d_3x$$

· · · · · · · ·

We conclude that d is dependent on c, i.e. c is relevant for d.

Marc-Wogau closes his discussion of the methods of variation with the observation that the second method

> is of great significance for establishing a functional relation of dependence between two factors is easy to see. What characterizes such a relationship is that one factor takes on different values when the values of the other vary. The method of variation is also of great importance for establishing a statistical correlation between two factors. (Marc-Wogau, 1950, pp. 192-193)

1.5 Supervenience and dependence

During the last decade or so the notion of supervenience has attracted a good deal of interest in the philosophical literature. The notion is used in theses like "the mental is supervenient on the physical" and "moral properties are supervenient on non-moral properties". How the term "supervenient" should be understood here is not clear and this has been much discussed. It seems to be relatively uncontroversial that supervenience is connected with the notion of dependence. The following quotation from Bonevac (1988) p. 37 may illustrate this.

> The concept of supervenience relates closely to our ordinary idea of dependence. Indeed, its advocates take supervenience to be a precise expression of our intuitive notions of dependence and determination. Generally, one realm—of properties, facts, events,

sentences, or models—*supervenes* on another just in case the latter determines the former; just in case, that is, the constitution of the former realm is a function of the constitution of the latter. Sociology supervenes on psychology, for example, if the psychological facts, taken together, determine the sociological facts. And our ordinary macro-level discourse supervenes on microphysics if macro-level circumstances depend on, or are functions of, microphysical circumstances. Supervenience thus gives us a way to explicate the primacy of physics and the unity of science. We can say that all the sciences supervene, ultimately, on physics. That is to say, the physical facts determine all scientific facts.

Thus, supervenience relates closely to our ordinary notion of dependence. The question now arises of how supervenience is related to the notion of structural dependence of interest here. Note first that structural dependence is primarily a relation between variable quantities or aspects while supervenience seems to have a wider application. But even if we disregard this difference it seems obvious that at least for some authors the connection between the two notions is not very strong, since for them supervenience lacks the property of being "undirected" which we take as characteristic of structural dependence (see section 1.3). However, even for the advocates of this view, the study of structural dependence could be of some use for the study of supervenience if supervenience entails some kind of structural dependence. For example, suppose that "Y is supervenient on X" does not mean the same thing as "Y is structurally dependent on X" but that the latter is—in some of the senses that will be ascribed to it—a necessary condition of the former. Then it would be interesting to find out exactly what kind of structural dependence holds between Y and X when Y is supervenient on X. However, this is not a problem area we shall enter into in the present study of structural dependence; the study of the relation between supervenience and structural dependence lies beyond the scope of this investigation.

1.6 Invariance and dependence

The mathematical notion of invariance is the most prominent formal device in our study of structural dependence. In this section we shall give a first, informal account of the notion—or perhaps better the general idea—of invariance as it is used in mathematics, and we shall also try to outline the connection between invariance and dependence.

In the beginning of the chapter about invariance in Bell (1945) the author emphasizes that "a comprehensive formal definition of invariance might be difficult to fabricate and unilluminating once it was constructed" (p. 420). Instead the author quotes the following informal description from C.J. Keyser, which he thinks gives the gist of the matter more intelligibly.

> Invariance is changelessness in the midst of change, permanence in a world of flux, the persistence of configurations that remain the same despite the swirl and stress of countless hosts of curious transformations. (Keyser, 1904, p. 313)

With this description in mind, it is not at all surprising that invariance is closely connected with structural dependence. *Invariant* means roughly *constant* or *unchanging,* and as we have seen earlier in this chapter, structural dependence has a lot to do with change when something else changes and constancy when something else is constant.

It is doubtful if one can talk about a single notion of invariance in mathematics. It is perhaps more adequate to say that there is a number of invariance notions, different in different branches of mathematics, but still related, possibly by some kind of family resemblance. This may be one of the reasons why it is difficult to fabricate a comprehensive formal definition of invariance. In Margenau (1972) there is nevertheless an attempt to characterize rather exhaustively the basic idea behind the different notions of invariance in an informal way. The following quotation contains a key passage.

> Strictly speaking, then, the idea of invariance remains incomplete unless it is coupled with a specification of the permitted changes or, if the changes are brought about by man, the permitted operations that are without influence on the quality asserted to be constant. Hence, in every meaningful statement about invariance, two things must be clearly set forth: (1) Whatever it is that remains invariant, (2) the changes, operations, or transformations under which invariance is said to hold....Let us speak of item 1 as the "invariant property" or entity and of item 2 as the "transformations" with respect to which invariance is asserted. (Margenau, 1972, p. 48)

Let us illustrate this characterization with a few examples. In Weyl (1949) two definitions of invariance is presented. On p. 9 Weyl says that

> $R(xy)$ is invariant with respect to the equivalence $\approx$, if $R(ab)$ always entails $R(a'b')$, provided $a' \approx a$ and $b' \approx b$.

On p. 73 he defines invariance as follows.

> A ternary relation $R(xyz)$ between points is *invariant* with respect to a given transformation $\sigma : p \rightarrow p'$ and its inverse $p' \rightarrow p$ if $R(abc)$ always implies $R(a'b'c')$ and vice versa.

That the relation in the first case is binary and in the second case ternary is of no importance. One difference between the two definitions concerns that with respect to which the relation is invariant. In the first definition invariance is understood with respect to an equivalence relation and in the second with respect to a transformation, i.e. a function or mapping. If we use Margenau's characterization we can say that the changes under which invariance is said to hold are substitutions of equal elements in the first definition and a transformation (in the sense of a mapping) in the second. We shall study the relation between these two definitions in detail in section 2.7.

The kind of invariance which will play the most important role in the present study of structural dependence is invariance of a relation with respect to a set of transformations. This notion is a slight extension of Weyl's second one. A relation ρ on a set A is invariant with respect to (under) a set Φ of one-to-one mappings of A onto itself if it is invariant with respect to every element φ in Φ. Thus, if ρ is n-ary, then ρ is invariant with respect to Φ when for all φ in Φ and all $a_1,...a_n$ in A

$$\rho(a_1,...,a_n) \text{ if and only if } \rho(\varphi(a_1),..,\varphi(a_n)).$$

The notion of automorphism invariance plays an important role in the sequel. Let ρ be a relation on the set A. According to standard terminology a one-to-one mapping or transformation φ of A such that for all $a_1,...a_n$ in A

$$\rho(a_1,...a_n) \text{ if and only if } \rho(\varphi(a_1),...,\varphi(a_n))$$

is called an automorphism of ρ. (An automorphism of ρ is thus an isomorphism from ρ to ρ.) The set of automorphisms of ρ is thus the set of one-to-one mappings or transformations of A relative to which ρ is invariant. Suppose that π is a relation on A. That π is automorphism-invariant relative to ρ means that π is invariant with respect to the set of automorphisms of ρ. It is clear after a moment's reflection that π is automorphism-invariant relative to ρ if and only of the set of automorphisms of ρ is a subset of the set of automorphisms of π. The notion of automorphism invariance will be used in the next section.

A great achievement in the mathematical study of invariance is the so called Erlangen Program. In 1872 Felix Klein was appointed professor of mathematics at the University of Erlangen. In his inauguration lecture, published as Klein (1872) and in Klein (1921) pp. 460-497, he suggested a classification of geometry using the mathematical notion of a group. According to Klein, for each geometry there is a group of transformations which characterizes it in the sense that the geometry is concerned with that which is invariant under this group. Klein's idea was met with an overwhelming response by the mathematical community and has become well-know under the name of the Erlangen Program. It has been applied not only to geometry but also to other parts of mathematics and even to disciplines outside mathematics, notably mechanics. The present investigation of structural dependence is within the general spirit of the Erlangen Program.

In modern terminology the Erlangen Program can be summarized as follows (see for example Kunle & Fladt, 1974). Let there be given a set M. The one-to-one mappings of M onto itself, i.e. the set of bijections or transformations of M, under functional compositions form a group. This group is often called the *symmetric group* S_M of M. Every subgroup of S_M is called a *transformation group* of M. Subsets of M are called *point-sets* or *figures*. Figures may have various kinds of properties. Those properties of figures that are left invariant under all transformations of a given transformation group are of special interest. As Klein himself has put it (in the translation taken from Kunle & Fladt, 1974, p. 462):

> Let there be given a manifold, and in it a group of transformations; it is our task to investigate those properties of a figure belonging to the manifold that are not changed by the transformations of the group.

A property $\mathbb{P}$ of a figure F is invariant under Φ if and only if $\mathbb{P}$ is a property of each figure $\varphi[F]$ for every φ in Φ, where $\varphi[F]$ is the figure into which F is mapped by the transformation φ (i.e. $\varphi[F] = \{\varphi(x) \mid x \in F\}$). The study of those properties of figures of M that are left invariant under all transformations of a given transformation group Φ is called a geometry and we denote it by (M,Φ). Note that if $\Psi \subseteq \Phi$ then what is left invariant under Φ is also left invariant under Ψ.

We can now define what it means for two figures to be equivalent or congruent within the frame of a geometry. Two figures F and F' of M are equivalent or congruent in the geometry (M,Φ) if there is a transformation φ in Φ such that φ takes F into F' (i.e. maps F to F').

Let us give a few examples of geometries. We let M be the two-dimensional plane. A transformation of the plane (i.e. a one-to-one mapping of M onto M) which preserves distances is called a *rigid mapping* (or rigid movement or isometry). φ is thus a rigid mapping if φ is a transformation of M and the distance between arbitrary points x and y is the same as the distance between $\varphi(x)$ and $\varphi(y)$. A transformation of the plane which changes all distances in a fixed scale is called a *similarity.* A similarity thus preserves ratios between distances. A one-to-one mapping of the plane which preserves parallelism and the ratios of parallel segments is called an *affine mapping*.

The set of all rigid mappings under functional composition form a group and the situation is analogous with the set of all similarities and the set of all affine mappings. Note that rigid mappings leave distances invariant , the similarities leave invariant ratios of distances, i.e. ratios of segments, whereas affine mappings leave invariant ratios of parallel segments. This implies that the group of rigid mappings is a subset of the group of similarities which is a subset of the group of affine mappings.

The geometry of rigid mappings, that is the geometry (M, Φ) where M is the plane and Φ the group of rigid mappings, is the Euclidean (metric) geometry. The geometry of similarities is called the similarity or equiform geometry. The geometry of affine mappings is called the affine geometry. Because of the relation between the three groups stated above, a theorem in affine geometry is also a theorem in equiform geometry and a theorem in equiform geometry also holds in Euclidean geometry. Distances are objects under study in Euclidean geometry but distance has no place in the equiform or affine geometry. Ratios between distances are studied in the Euclidean and equiform geometry but not in affine geometry.

The Erlangen Program guided geometrical research for almost fifty years. The basic idea behind the program is the study of invariants under transformations and this conception has had a great influence in many branches of mathematics. But not only in mathematics; it has been applied in other subjects too, notably in physics. Klein himself wrote in 1912 (the quotation taken from Blumenthal and Menger, 1970, p. 26):

> What the modern physicists call "relativity theory" is the theory of the invariants of a four-dimensional space-time continuum (Minkowski's world) with respect to a given group of collineations (the Lorentz group), and hence is a geometry.

Another area in which the spirit of the Erlangen Program can be clearly seen at work is in S.S. Stevens' classification of scale types. For an informal presentation Stevens' own words from a rather late paper serve nicely.

> ...a scale type is defined by the group of transformations under which the scale form remains invariant, as follows.
> A *nominal scale* admits any one-to-one substitution of assigned numbers. Example of a nominal scale: the numbering of football players.

> An *ordinal scale* can be transformed by an increasing monotonic function. Example of
> an ordinal scale: the hardness scale determined by the ability of one mineral to scratch
> another.
> An *interval scale* can be subjected to a linear transformation. Examples of interval
> scales: temperature Fahrenheit and Celsius, calendar time, potential energy.
> A *ratio scale* admits only multiplication by a constant. Examples of ratio scales: length,
> weight, density, temperature Kelvin, time intervals, loudness in sones. ...
> The permissible transformations defining a scale type are those that keep intact the
> empirical information depicted by the scale. If the empirical information has been preserved,
> the scale form is said to remain invariant. (Stevens, 1968, p. 850)

In this quotation from Stevens he defines the scale types in terms of groups of
transformations of the real numbers. (Stevens does not seem to have been quite
consistent on this point in all his writings on scale types.) The group that defines
the nominal scale type consists of any one-to-one transformation and Stevens calls it
the permutation group or the symmetric group; the second name is preferred here.
The ordinal scale type is defined by the group of all monotonic increasing functions
and is called the isotonic group. The group defining the interval scale type is the
affine or linear group consisting of all transformation of the form x'=ax+b.
Finally, the similarity group consisting of all transformations of the form x'=ax
defines the ratio scale type. Note that the similarity group is a subgroup of the
linear group, which is a subgroup of the isotonic group, and the isotonic group is of
course a subgroup of the symmetric group.

An essential part of Stevens theory of scale types is the following idea: which
statistical operations are appropriate to use on data depends on the scale type of the
data. The criterion for appropriateness of a statistic with respect to a scale type is
according to Stevens *invariance* under the group of transformations which define
the scale type.

> Thus the case that stands at the median (midpoint) of a distribution maintains its 'middleness'
> under all transformations that preserve order (isotonic group), but an item located at the mean
> remains at the mean (retains its 'meanness'!) only under transformations as restricted as
> those of the linear group. The ratio expressed by the coefficient of variation remains
> invariant only under the similarity transformation (multiplication by a constant). (Stevens,
> 1951, p. 24)

Median is therefore an appropriate—or permissible as Stevens also says—statistical
operation for ordinal scales, while mean is permissible for interval scales and
coefficient of variation for ratio scales. The classes of permissible statistics for the
four scale types constitute a cumulative series; what is permissible for nominal
scales is also permissible for ordinal scales and so on. This follows immediately

from the inclusion relations which hold between the defining groups.

The analogy between the Erlangen Program and Stevens' theory of scale types is rather obvious. According to the Erlangen Program we use groups of transformations to define geometries. In a given geometry we study those properties of figures that are left invariant under the defining group of transformations. According to Stevens we use groups of transformations to define scale types. In the study of data expressed in a given scale type we use statistical operations that are invariant under the group of transformations which defines the actual scale type. It was noted above in the discussion of the Erlangen Program that ratios between distances are studied in the Euclidean and equiform geometry but not in the affine geometry. In the remarks about Stevens' classification of scale types we pointed out that mean is a permissible statistical operation for ratio and interval scales but not for ordinal scales.

The notion of invariance has often been connected with objectivity and meaningfulness. This can be illustrated with the following two quotations from Hermann Weyl. (The italics are added here.)

> A relation between world-points has an *objective meaning* if, and only if, it is defined by such arithmetical relations between coordinates of the points as are invariant with respect to the transformations (III) [i.e. Galilei transformations]. (Weyl, 1922, p. 155)

> We start with a group Γ of transformations. It describes, as it were, to what degree our point field is homogeneous. Once the group is given we know what likeness or similarity means—namely two figures are similar (or alike, or equivalent) that arise from each other by a transformation of Γ—and also under what condition a relation is *objective*, namely if it is invariant with respect to all transformations of Γ. It is in this sense that Felix Klein in his famous Erlanger Program (1872) promulgated the conception that a geometry is determined by a group of transformations. (Weyl, 1949, pp. 73-74.)

Patrick Suppes is perhaps the philosopher who has most strongly emphasized the importance of invariance for the notion of meaningfulness. Among other things Suppes has made it clear that Stevens's conception of permissible statistics is a special case of a more general question of meaningfulness, namely which numerical statements containing measure values express something empirically meaningful. Suppes' answer, as given in Suppes and Zinnes (1963) p. 66, is the following:

> A numerical statement is *meaningful* if and only if its truth (or falsity) is constant under admissible scale transformations of any of its numerical assignments, that is, any of its numerical functions expressing the results of measurement.

This kind of meaningfulness consists, therefore, of invariance of truth values under admissible scale transformations.

The interest among philosophers in the mathematical notion of invariance seems to be increasing. In this essay, we shall use the notion of invariance in the investigation of the notions of dependence and independence. The close connection between invariance and structural (in)dependence can be illustrated with an English translation of a short passage from Klein's Erlangen Program, the translation taken from Gårding (1977) p. 102.

> There are transformations of [ordinary space] leaving invariant geometric properties of space configurations. In fact, geometric properties are in themselves independent of position, absolute magnitude and orientation of the object under consideration. The properties of a space configuration do not change under movements in space, similarity transformations and reflections and all transformations generated by them. The totality of all these transformations we shall call the *capital group* of space transformations: *geometric properties are invariant under the transformations of the capital group*. This can be turned around: *geometric properties are characterized by being invariant under transformations of the capital group*.[1]

That part of Klein's idea in which we are interested for the moment may be understood as follows. Geometric properties of space configurations are independent of position, absolute magnitude and orientation of the object under consideration. Hence, if we change only position, absolute magnitude and orientation the geometric properties (of space configurations) are unchanged. The change of only position, absolute magnitude and orientation is the result of a transformation of the capital group. Therefore, geometric properties of space configurations are invariant under transformations of the capital group.

Klein's way of reasoning is in accordance with the first method of variation used

---

[1]The original German text runs as follows:
Es gibt nun räumliche Transformationen, welche die geometrischen Eigenschaften räumlicher Gebilde überhaupt ungeändert lassen. Geometrische Eigenschaften sind nämlich ihrem Begriffe nach unabhängig von der Lage, die das zu untersuchende Gebilde im Raume einnimmt, von seiner absoluten Grösse, endlich auch von dem Sinne, in welchem seine Teile geordnet sind. Die Eigenschaften eines räumlichen Gebildes bleiben also ungeändert durch alle Bewegungen des Raumes, durch seine Ähnlichkeitstransformationen, durch den Prozess der Spiegelung, sowie durch alle Transformationen, die sich aus diesen zusammensetzen. Den Inbegriff aller dieser Transformationen bezeichnen wir als die Hauptgruppe räumlicher Änderungen; *geometrische Eigenschaften werden durch die Transformationen der Hauptgruppe nicht geändert*. Auch umgekehrt kann man sagen: *Geometrische Eigenschaften sind durch ihre Unveränderlichkeit gegenüber den Transformationen der Hauptgruppe charakterisiert*. (Klein, 1921, p. 463.)

for determining independence (i.e. irrelevance)—see the first method of variation in the quotation from Marc-Wogau in section 1.4. It seems therefore obvious that when Klein says that geometric properties are independent of position, absolute magnitude and orientation he means that position, absolute magnitude and orientation are not relevant for geometric properties (of space configurations).

The connection between irrelevance and invariance can be described in a preliminary way as follows: α is irrelevant for β if [the structure of] β is invariant under transformations that change only [the structure of] α. As we shall see, invariance is the central notion even for the explication of determination, but then invariance is used in another way: β is determined by α if [the structure of] β is invariant under transformations that preserve [the structure of] α.

1.7 Independence of primitive symbols

It is dependence and independence between aspects or variable quantities that are mainly of interest in this study. But there is a notion of independence found in modern logic which will play a crucial role in the sequel, and this is independence of primitive symbols. Let us take a look at this.

Suppose P, Q and R are the primitive symbols of a theory $\mathbb{T}$. P is then said to be *independent* of Q and R if it is not possible to define P in terms of Q and R. A standard procedure to show that P is independent of Q and R is the following: Find two models M_1 and M_2 of $\mathbb{T}$ which have the same domain and such that the interpretation of Q is the same in both models and analogously for R, *but* such that P is interpreted differently in M_1 and M_2. This is known as Padoa's method, named after the Italian logician Allessandro Padoa who formulated it in 1900. Padoa's method gives a sufficient condition for a symbol to be independent of the remaining primitives of a theory.

Padoa's method is intimately related to the idea of invariance. In fact, Padoa's method can be formulated as follows. To show that P is independent of Q and R find two models $M_1 = \langle A_1, \pi_1, \omega_1, \rho_1 \rangle$ and $M_2 = \langle A_2, \pi_2, \omega_2, \rho_2 \rangle$ of $\mathbb{T}$, where π_i, ω_i and ρ_i (i=1,2) are the interpretations of P, Q and R, respectively, such that there is an isomorphism φ from $\langle A_1, \omega_1, \rho_1 \rangle$ to $\langle A_2, \omega_2, \rho_2 \rangle$ which is not an isomorphism from $\langle A_1, \pi_1 \rangle$ to $\langle A_2, \pi_2 \rangle$. If we use $\mathbb{I}(\mathbf{X}, \mathbf{Y})$ to denote the set of isomorphisms of the structure $\mathbf{X}$ to the structure $\mathbf{Y}$ we can say that Padoa's method amounts to the following: To show that P is independent of Q and R find two models $M_1 = $ $= \langle A_1, \pi_1, \omega_1, \rho_1 \rangle$ and $M_2 = \langle A_2, \pi_2, \omega_2, \rho_2 \rangle$ of $\mathbb{T}$, where π_i, ω_i and ρ_i (i=1,2) are the interpretations of P, Q and R, respectively, such that it is not the case that

$$I(\langle A_1,\pi_1\rangle,\langle A_2,\pi_2\rangle) \supseteq I(\langle A_1,\omega_1,\rho_1\rangle,\langle A_2,\omega_2,\rho_2\rangle).$$

The connection between Padoa's method and the concept of invariance is perhaps made clearer by the observation that a relation π on A is automorphism-invariant relative to $\langle A,\omega,\rho\rangle$ if and only if

$$I(\langle A,\pi\rangle,\langle A,\pi\rangle) \supseteq I(\langle A,\omega,\rho\rangle,\langle A,\omega,\rho\rangle) .$$

(The notion of automorphism invariance was introduced in the preceding section.)

In 1953 E.W. Beth showed that for first-order theories Padoa's method also gives a necessary condition for a symbol to be independent of the remaining primitives. Somewhat more precisely Beth's theorem applied to the first-order theory T with the primitive symbols P, Q and R says that P is definable in terms of Q and R, i.e. P is dependent on Q and R, if and only if for every pair of models $M_1 = \langle A_1,\pi_1,\omega_1,\rho_1\rangle$ and $M_2 = \langle A_2,\pi_2,\omega_2,\rho_2\rangle$ of T, where π_i,ω_i and ρ_i (i=1,2) are the interpretations of P, Q and R, respectively, it holds that

$$I(\langle A_1,\pi_1\rangle,\langle A_2,\pi_2\rangle) \supseteq I(\langle A_1,\omega_1,\rho_1\rangle,\langle A_2,\omega_2,\rho_2\rangle).$$

Beth's theorem thus connects definability in first-order theories with the algebraic idea of invariance. Since definability means (definitional) dependence Padoa's method and Beth's theorem provide a bridge between dependence and invariance. And this bridge is one of the roads we shall follow in this study. For we shall let the condition of algebraic invariance in Padoa's method and Beth's theorem define a notion of dependence. But we shall not apply this concept to symbols but to systems. And we shall regard this notion as meaning complete dependence; if the invariance condition defining the concept is not fulfilled, this does not imply "independence" but only "not complete dependence"—a weaker form of dependence is still possible.

Chapter 3 and the first part of chapter 7 is devoted to a characterization of the kind of system to which the notions of dependence and independence will be applied. For the moment we just remark that these systems have a relational character and can be used to formulate what are often called attributes, aspects, qualities, quantities and so on. The next section contains a first presentation of the basic concept of a relation in terms of which the systems are characterized. This notion of a relation differs somewhat from the traditional one.

1.8 Relations as functions

To be able to study dependence between aspects it is necessary to have a rather precise view of what an aspect is. It is common in modern measurement theory to regard an aspect as (or characterized by) a relational structure, i.e. an ordered tuple consisting of a set with one or more relations on this set. A relation is usually regarded as a set of ordered n-tuples, where n is a positive integer. I shall depart from this tradition in this study, and make use of the different but nonetheless equally natural idea that the extension of a relation over a set can be represented as a relation in the ordinary set-theoretical sense of a set of ordered n-tuples. The question what a relation is in itself I will leave open; it is not important for the purpose of this study. But it is my fundamental hypothesis that the mathematical form of a relation is a function that has sets as arguments and relations in the usual set-theoretical sense as values. As will be seen, the difference between this approach and the traditional one is not great; I think it is fair to say that I make explicit what is more or less implicit in the usual presentation of measurement theory.

It is important here to make clear that it is relations connected with aspects, for example *equality* and *greater than* in different respects, that I will regard as functions. In the study of these relations I shall use other relations which are treated in the usual set-theoretical way as sets of ordered n-tuples. It is therefore convenient to have a simple term to denote a relation regarded as a function to reduce the risk of confusion. After some hesitation I have decided on *relational*. As a noun I believe this is a new word, and the reason that I dare to introduce it is the existence of the noun *functional* for a special kind of function in recursion theory. The noun relational is thus constructed by analogy with the noun functional, but that does not imply that the relationship between a relation and a relational is at all similar to the relationship between a function and a functional. A relational is a relation regarded as a function of a special kind. If we look upon a set-theoretical relation as a representation of a relation in the ordinary language sense of the word, then a relational is another representation of such a relation, namely as a function of a special kind.

I will regard aspects as systems of relationals and such systems I will often call relational systems. I shall therefore distinguish between a relational system and a relational structure. A relational structure is understood here in the usual sense of the term, namely an ordered sequence consisting of a set and relations on this set, where the relations are sets of ordered n-tuples.

It is worth emphasizing perhaps that even though I do not treat relations in connection with aspects as sets of ordered n-tuples, but as relationals, I work to a large extent within the ordinary set-theoretical framework: a value of a relational is a set-theoretical entity. Suppose R is a relational. Then R(A) is the graph or extension of R on A and states what holds for A with respect to R. Let us suppose for simplicity that R is binary. $\langle x,y \rangle \in R(A)$ means that R holds between x and y in A. Note that R(A) is a set of ordered pairs.

One result of treating relations as functions is that it becomes meaningful to ask if a relation is stable in the following sense: If x_1 and x_2 belong to both A and B then R holds between x_1 and x_2 in A if and only if R holds between x_1 and x_2 in B. As we shall see in chapter 5, this stability condition is, in the context of social choice, closely related to Arrow's famous condition of independence of irrelevant alternatives. Note that the term "independence" here is used in a sense distinct from those we are interested in; it is not related to the notion of dependence or independence between aspects.

Chapter 3 is devoted to developing the theory of relationals, and in chapter 7 this theory is extended to systems of relationals. It is to relationals and systems of these that the notions of dependence and independence will be applied.

1.9 Notions of independence in modern measurement and decision theory

In modern measurement and decision theory several notions of dependence and independence are used. In this section I shall try to outline some of these notions. Perhaps the most important notion of independence not dealt with is the notion of independent events in probability theory. This notion falls outside the scope of this study. We shall instead restrict ourselves to notions of dependence and independence used in connection with conjoint measurement and multidimensional decisions. In this contexts the focus is on independence rather than dependence.

Chapter 6 in Krantz et.al.(1971) deals with additive conjoint measurement. Section 1 of this chapter has the title "Several notions of independence". The first notion of independence presented there is the independence of variables or, as Krantz et.al. prefer to call it, *independent realizability* of components (see p. 246). Krantz et.al. avoid the "variable" terminology because, among other reasons, it suggests that numerical representations already exist, and they do not want to assume this. Two components A_1 and A_2 are *independently realizable* if for every a in A_1 and p in A_2, $\langle a,p \rangle$ must be a realizable entity. This means that the value for each component can be chosen without regard to the value for the other.

Krantz et.al. (1971) state two quite different reasons why the postulate of independent realizability might fail. One of these—that of the greatest interest here—is the following: an empirical law relates the two components. To explain it they give a familiar example, viz. density.

> Suppose that we have a homogeneous substance and let A_1 denote the continuum of possible masses and A_2, the continuum of possible volumes. Once the mass has been selected, then under fixed experimental conditions only one volume can occur. That is, mass and volume are not independent components for a given substance since m=dV, where d is known as the density of the substance. Contrast this with momentum where, in principle, any velocity may be paired with any mass. (Krantz et. al., 1971, pp. 246-247)

The idea of independent realizability will play an important role in the sequel.

The second notion of independence treated in section 6.1 of Krantz et.al. (1971) (see p. 247) is that the two components contribute their effects independently to the attribute in question. This holds when there are numerical scales on the two components and a rule for combining them such that the resultant measure preserves the qualitative ordering of the attribute, and we then say that the attribute is decomposable. More precisely, the structure $\langle A_1 \times A_2, \succeq \rangle$ is *decomposable* if there are real-valued functions φ_i on A_i, i=1,2 and a function F from Re×Re into Re, 1:1 in each variable, such that, for all a,b in A_1 and p,q in A_2

$$\langle a,p \rangle \succeq \langle b,q \rangle \text{ iff } F(\varphi_1(a),\varphi_2(p)) \geq F(\varphi_1(b),\varphi_2(q)). \tag{1.10.1}$$

($\langle A_1 \times A_2, \succeq \rangle$ thus represents the attribute and A_1 and A_2 the components.) In a decomposable structure the components contribute their effects independently.

If the combining rule F is such that

$$F(x,y) = x+y$$

then we have a special kind of independence, viz. additive independence. In this case equation (1.10.1) above establishes a measure of the attribute in which the components contribute to it additively independently.

Another notion of independence in section 6.1 of Krantz et.al. (1971) is that of independent relations (p. 248f.). A binary relation $\succeq$ on $A_1 \times A_2$ is *independent* iff
(1) for a,b∈ A_1, $\langle a,p \rangle \succeq \langle b,p \rangle$ for some p∈ A_2 implies that $\langle a,q \rangle \succeq \langle b,q \rangle$ for every q∈ A_2,

(2) for $p,q \in A_2$, $\langle a,p \rangle \gtrsim \langle a,q \rangle$ for some $a \in A_1$ implies that $\langle b,p \rangle \gtrsim \langle b,q \rangle$ for every $b \in A_1$.

The term "independent" is used here because interaction between the components is ruled out. (See Luce & Suppes, 1965, p. 269.)

An independent relation induces a natural ordering $\gtrsim_i$ on each component in the following way.

(i) For $a,b \in A_1$, $a \gtrsim_1 b$ iff for some $p \in A_2$, $\langle a,p \rangle \gtrsim \langle b,p \rangle$.

(ii) For $p,q \in A_2$, $p \gtrsim_2 q$ iff for some $a \in A_1$, $\langle a,p \rangle \gtrsim \langle a,q \rangle$.

The notion of an independent relation is connected with the notion of a decomposable structure as follows. For any decomposable structure $\langle A_1 \times A_2, \gtrsim \rangle$, if F in equivalence (1.10.1) is strictly monotonic in each variable, the relation $\gtrsim$ is independent. (See Krantz et. al., 1971, p. 250). If F is such that $F(x,y)=x+y$ then F is of course monotonic in each variable and $\gtrsim$ is therefore independent.

The idea of independence behind the definition of a decomposable structure is that the components contribute their effects to the attribute in question independently. This is also the case with the definition of an independent relation. The concept of independent realizability is, however, of a different character. It implies that the components are totally unconnected. The notion of independence we are interested in in this study is more related to independent realizability than to decomposability and independent relations.

In multidimensional decision theory there is an independence condition of great importance, called "utility independence" (in Roberts, 1979, "strong independence"). It is a generalization of the condition of independence for relations mentioned above and has a lot to do with additive utility functions. The idea behind this independence condition is simple. Suppose $\gtrsim$ is your preference relation over a set A of alternatives. Suppose further that A can be regarded as the Cartesian product of the sets A_1 and A_2, i.e. $A=A_1 \times A_2$. Let us now suppose that you are indifferent between the alternative $\langle x,a_2 \rangle$ and a lottery which gives $\langle a_1,a_2 \rangle$ with probability 1/2 and $\langle b_1,a_2 \rangle$ with probability 1/2. The situation can be illustrated as follows:

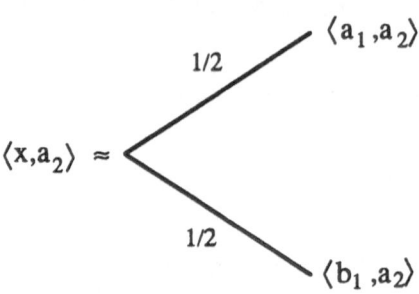

Suppose now that in the above example the second alternative is changed to some other element b_2, so you have a choice between the alternative $\langle x,b_2 \rangle$ and a lottery which gives you $\langle a_1,b_2 \rangle$ with the probability 1/2 and $\langle b_1,b_2 \rangle$ with the probability 1/2. Will it change your preferences or will you still be indifferent between the alternative and the lottery as the illustration below indicate?

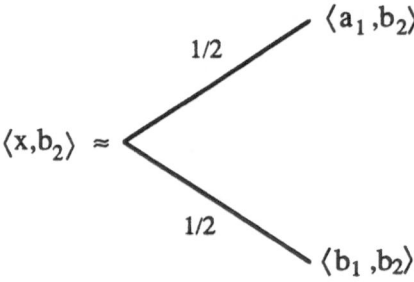

If the answer is that you will still be indifferent, your actual preferences will not depend on the fixed second coordinate. If, in a choice between an alternative and a lottery whose consequences have the same second coordinate as the alternative, your preferences do not depend on the second coordinate, then for you the first component A_1 is utility independent of the second A_2. This is of course not an exact definition of utility independence, but only a rough sketch. However, I hope it will suffice to show that this notion of independence resembles those of decomposability and independent relations in dealing with the independence of (or non-interaction between) the contribution of each component. An extensive discussion of utility independence is to be found in, for example, Keeney & Raiffa (1976) and Roberts (1979).

1.10 Applications of structural dependence

The aim of this investigation is to study structural dependence between aspects, both in the sense of determination and in the sense of relevance. One important task is to try to characterize the different degrees of strength in which dependence can be found. The hope is to construct some kind of theory of structural dependence. I think it important to point out that the theory of structural dependence I am striving towards does not constitute an analysis of a single concept already existing in scientific language. I like to think of this investigation of structural dependence as an attempt at constructing tools which can function as means of analysis.

At this point it is natural to ask in which contexts the theory of structural dependence can be applied. Most natural and social sciences study to some degree connections, correlations or relationships between factors or aspects—in other words they study dependence between variables. Such a study is often said to be "quantitative" and involves applications of "quantitative methods". Dependence between variables is therefore a central problem area for "quantitative methodology" and a theory of structural dependence has a role to play here.

Quantitative relationships between aspects are often described in terms of numbers and are therefore expressed by means of scales or measures. The study of scales and measures is the subject matter of measurement theory. The theoretical study of quantitative relationships belongs therefore to a large extent to measurement theory. This implies that the theory of structural dependence is chiefly applied to the natural and social sciences as a part of measurement theory.

However, I think that the theory of structural dependence can also be applied to certain problem areas in measurement theory itself. To be more specific, I believe that the theory of structural dependence is of interest to the meaningfulness problem and to the study of the form of numerical laws (especially the problem of dimensional invariance). Even for the theoretical study of decomposing complex phenomena into factors, which is important in measurement theory, and for the problem of aggregating or amalgamating factors into a more compound aspect which is central in multidimensional measurement and decision theory, the theory of structural dependence seems to be useful.

The relation between the theory of structural dependence and the different areas of measurement theory mentioned above can perhaps be illustrated in the following way.

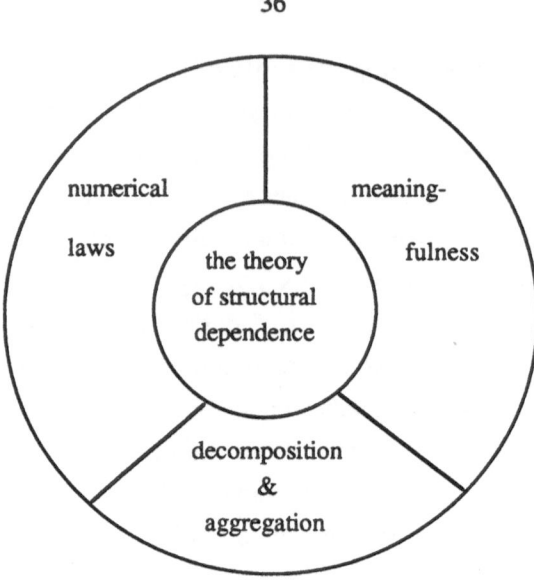

1.11 Summing up

In this chapter I have tried to outline and delimit the subject of this investigation in an informal way. The presentation has therefore been preliminary and incomplete at many points, and I have avoided a lot of distinctions which I regard as necessary. In the rest of the book I shall be more formal, using the language of elementary set theory, algebra and logic.

Let me end this chapter with a summary of what I think is important as a starting-point for the rest of this book. The subject matter of the investigation is structural dependence and independence between aspects. This is, as far as I can see, one—or perhaps *the*—cluster of dependence notions in applications of mathematics. A distinction between dependence in the sense of determination and in the sense of relevance is made. To study determination we use the idea of preservation of equality applied to two different "levels", the object level—which gives rise to the notion of conformity—and the structure level—which gives rise to the notion of subordination. See the figure below. The notion of relevance will be studied in terms of the notion of determination.

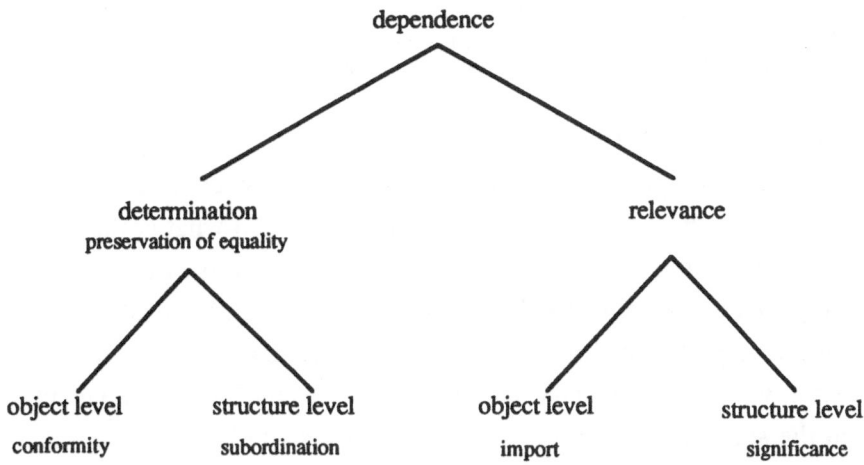

Structural dependence is strongly connected to the idea of concomitant variations and to the mathematical notion of invariance. This implies that the study of structural dependence has roots going back to Mill's inductive logic and to Klein's Erlangen Program for geometry. Padoa's method and Beth's theorem on definability are also important starting-points. Structural dependence is related, in addition to the notions of concomitant variations and invariance, even to objectivity and meaningfulness.

One of the reasons for the attribute "structural" in "structural dependence" is to emphasize that the notion of dependence which will be dealt with here has an "undirected" character. Dependence in the sense of determination is for example symmetric. Structural dependence must therefore be distinguished from other kinds of dependence which are not "undirected", for example causal dependence.

The concept of dependence we are interested in is closely related to the concept of a function in mathematics. It is probably in the phrase "is a function of" that the connection between dependence and function finds its strongest expression. As Edmund C. Berkeley has put it:

> When something *depends on* something else, as we say in common everyday talk, then that something *is a function of* the something else, to use the language of mathematics. ...
> Whenever something Y (a decision, a condition, a magnitude, a thing, a person, etc.) *depends on* one or more other factors or elements X (decisions, conditions, magnitudes, things, persons, etc.), then Y is said to be *a function of* X. In other words, Y is said to be a function of X if when X is given, Y is determined. Notice that here the symbol X may stand for just one factor or element or may stand for two or more; the symbol X, in other words, may be interpreted as singular or plural. The word "determined" means "settled",

"fixed", "found out", "established"; the "determining" may involve a connection of cause and effect, or may refer to some very different kind of association or correlation where no cause and effect are involved at all. (Berkeley, 1966, p. 146-147)

The dependence concept is thus closely connected with the function concept. But as I have tried to indicate previously in this chapter the connection is not entirely straightforward. *To be a function of* is a relation between what perhaps could be called variable quantities. Dependence holds between factors or aspects. It is therefore important in what sense a factor or an aspect can be regarded as a variable quantity. Aspects as systems of relationals (i.e. systems of relations-as-functions) is the suggestion which will be elaborated here. As we shall see, the concept of a function in the sense of a univalent correspondence will play a predominant role in our study of dependence between systems of relationals. I think it is correct to say that this essay is an investigation of some aspects of the functional idea of dependence.

The reason for using the term "structural dependence" rather than "functional dependence" is that functional dependence is a more specific notion; functional dependence means complete determination. Structural dependence can, however, be divided into determination and relevance which can both be found in degrees from complete determination and relevance to complete undetermination and irrelevance. The main part of the investigation is devoted to a study of these "scales" of determination and relevance. The most basic concept, the "atom" of the study, is complete determination, which "on the structure level" will be called subordination and is characterized in terms of the algebraic notion of invariance. Subordination will be used in the investigation of partial determination and in the explication of complete relevance. This implies that the notion of invariance plays a considerable role in the study of structural dependence. This study will therefore be mainly algebraic in character.

Subordination is also closely related to explicit definability. But we will not go deeply into the theory of definition since we will avoid taking the language we use in the treatment of aspects into explicit consideration as a formal language. This is in conformity with what is done in algebra, where much of the theory can be developed without formalizing the language used.

Aspects will be characterized in terms of relations, more specifically as relational systems. But by "relation" here is not meant relation in the ordinary sense of a set of ordered n-tuples. Rather a relation is a function which takes sets as arguments and sets of ordered n-tuples as values. Such a relation will be called a relational. Some of the notions of independence in the measurement theoretic literature are closely related to the notions of independence investigated here, for example independent realizability.

CHAPTER 2

BASIC FORMAL CONCEPTS AND TERMINOLOGY

2.0   Introduction

This chapter is devoted to, on the one hand, a brief presentation of the standard
logical and mathematical terminology and theory which will be frequently used in
the rest of this book, and on the other, to proofs of some results within elementary
set theory and algebra which will be used in later chapters. It is intended to
function mainly as a kind of "dictionary" which the reader can consult when
necessary. Section 2.1 and the beginning of sections 2.7 and 2.9 contain the basic
apparatus for Part 2 and should therefore be at least cursorily perused before
proceeding further. But the remaining sections can wait until they are referred to.
Even though most terms which will be used are defined, a familiarity with elementary set theory and logic as presented in for example Suppes (1957) or Stoll
(1963) is presupposed.

"Iff" is frequently used as an abbreviation for "if and only if", both in this
chapter and in the rest of the book. "♦" marks the end of a proof. The existential
quantifier is written "∃" and the universal quantifier "∀".

Theorems, lemmas, corollaries and some equations are numbered consecutively.
"2.10.3", for example, refers to the third theorem, lemma, corollary or equation
(whatever it may be) found in section 2.10.

2.1   Relations and functions

In this section some basic facts and terminology pertaining to the set theoretical
notions of a relation and a function are presented.

An *n-ary relation* ρ (where n is a positive integer) in a set A (I sometimes say "on" A) is a subset of the Cartesian product A^n. If n=1 then ρ is called *unary*, if n=2 *binary* and if n=3 *ternary*.

Suppose ρ is a binary relation in A. $\langle a,b \rangle \in \rho$ will often be written $\rho(a,b)$ or $a\rho b$. The *domain* of ρ is $\{a \in A \mid \exists b : a\rho b\}$ and we denote it $Do\rho$. The *range* of ρ is $\{b \in A \mid \exists a : a\rho b\}$ and we denote it $Rg\rho$. The *field* of ρ is the union of the domain and the range of ρ.

If ρ is a binary relation, then the *converse* of ρ is the relation ρ^c defined by

$$a\rho^c b \ \text{ iff } \ b\rho a.$$

Hence,

$$\rho^c = \{\langle a,b \rangle \mid \langle b,c \rangle \in \rho\}$$

A binary relation such that no two distinct members have the same first coordinate is called a *function*. Thus, ρ is a function iff whenever $\langle a,b \rangle \in \rho$ and $\langle a,c \rangle \in \rho$ then b=c. If f is a function then we often write f(a)=b rather than $\langle a,b \rangle \in f$.

Suppose f is a function. If A is the domain of f then f is a function *on* A. If A includes the domain of f then f is a function *from* A. If B is the range of f then f is a function *onto* B. If B includes the range of f then f is a function *into* B. If f is a subset of A×B then f is thus a function from A into B. And if A is the domain of f and B is the range of f, then f is a function on A onto B. That f is a function on A into B is often denoted f:A→B.

Every function is a binary relation. Therefore, the converse f^c of f is well-defined. If also f^c is a function, then it is customary in mathematics to say that f^c is the *inverse* of f and denote it by f^{-1} (i.e. use "-1" as superscript instead of "c"). The converse of a function f is a function iff f is one-to-one. Now and then in the literature the inverse terminology and the notation "-1" is extended even to relations. The converse of a relation ρ is then called the inverse of ρ and denoted ρ^{-1}. This usage will be adopted here together with the use of "converse" and "c". We thus have two ways of expressing the same thing. This may seem superfluous, but we shall adopt different terminology in different contexts, which will be explained further in section 3.1. For the moment we note that the converse of a binary relation ρ, ρ^c, and the inverse of ρ, ρ^{-1}, is the same thing, viz.

$$\{\langle a,b \rangle \mid b\rho a\}.$$

If ρ and σ are binary relations then the *relative product* $\rho \mid \sigma$ of ρ and σ is defined by

$$\rho \mid \sigma = \{ (x,y) : \text{for some } z, \ x\rho z \text{ and } z\sigma y \}.$$

If f and g are functions then the relative product of f and g is symbolized by $g \circ f$

and is called the *composition* of f and g. Thus $g o f = f \mid g$. We call the operation o
functional composition. Thus, functional composition is relative product applied to
functions. We shall in some contexts make a distinction between composition, o , on
one hand and relative product, $\mid$, on the other, for example in section 3.1.

If f:A→B and f is onto B we say that f is *surjective*. If f:A→B and the inverse
of f is a function, then we say that f is *injective*. If f:A→B and f is both surjective
and injective, then we say that f is *bijective* and call f a bijection. The set of all
one-to-one functions on A onto B we denote by $\mathbb{B}i(A,B)$, $\mathbb{B}i$ for bijections. The set
of all one-to-one functions on A onto A we denote for simplicity $\mathbb{B}i(A)$ rather than
$\mathbb{B}i(A,A)$. An element in $\mathbb{B}i(A)$ is called a bijection on (or of) A.

A function $f:A^n \to A$ is said to be an n-*ary operation in* A. Note that an n-ary
operation f in A is a subset of A^{n+1} and thus an n+1-ary relation in A.

The identity function on a set A is denoted by ι_A and defined by

$$\iota_A(a) = a$$

for all $a \in A$.

As is well-known, the set of bijections of a non-empty set together with func-
tional composition constitutes a group. In other words, $\langle \mathbb{B}i(A), o \rangle$ is a group if A is
non-empty. The unit element in this group is the identity function ι_A on A and the
inverse of an element $\varphi \in \mathbb{B}i(A)$ is φ^{-1}.

Let ρ be a n-ary relation in A. The restriction of ρ to B is $\rho \cap B^n$ and denoted
by ρ/B.

If f is a function from A into B then $f \lceil X$ is the set $\{\langle a,b \rangle \in f \mid a \in X\}$ which of
course is a function from X into B. If f is a function on A into B and $X \subseteq A$ then
$f \lceil X$ is a function on X into B. $f \lceil X$ is often read as "the restriction of f to X" which
however not should be confused with f/X.

2.2 Properties of binary relations

Suppose ρ is a binary relation in A. We list some definitions of well-known formal
properties of ρ.

ρ is *reflexive* in A iff $a\rho a$ for all a in A.

ρ is *irreflexive* in A iff $\neg a\rho a$ for all a in A.

ρ is *symmetric* in A iff $a\rho b$ implies $b\rho a$ for all a,b in A.

ρ is *asymmetric* in A iff $a\rho b$ implies $\neg b\rho a$ for all a,b in A.

ρ is *antisymmetric* in A iff $a\rho b$ and $b\rho a$ implies a=b, for all a,b in A.

ρ is *transitive* in A iff $a\rho b$ and $b\rho c$ implies $a\rho c$, for all a,b,c in A.

ρ is *negatively transitive* in A iff ¬aρb and ¬bρc implies ¬aρc for all a,b,c in A, or equivalently, iff aρb implies aρc or cρb for all a,b,c in A.

ρ is *strongly complete/strongly connected* in A iff aρb or bρa for all a,b in A.

ρ is *complete/connected* in A iff a≠b implies aρb or bρa for all a,b in A.

ρ is an *equivalence relation* in A iff ρ is reflexive in A, symmetric and transitive.

It is often convenient to consider the *relational structure* ⟨A,ρ⟩ where ρ is a relation in A and, despite whatever abuse of language may be involved, we shall often simply call ⟨A,ρ⟩ a relation. ⟨A,ρ⟩ is a reflexive relation iff ρ is reflexive in A, and ⟨A,ρ⟩ is symmetric iff ρ is symmetric, and analogously with the other properties.

2.3. Order relations

Let ρ be a binary relation in A. ⟨A,ρ⟩ is a

- *quasi order (preorder)* iff ρ is reflexive and transitive in A
- *strict quasi order (strict preorder)* iff ρ is irreflexive and transitive in A
- *partial order* iff ρ is reflexive, transitive and antisymmetric in A
- *strict partial order* iff ρ is transitive and asymmetric in A
- *weak order* iff ρ is transitive and strongly complete in A
- *strict weak order* iff ρ is asymmetric and negatively transitive in A
- *simple / linear order* iff ρ is transitive, antisymmetric and strongly complete in A
- *strict simple order* iff ρ is transitive, asymmetric and complete in A.

A strict partial order is often defined as a relation which is irreflexive and transitive. (See for example Fishburn, 1970, p. 15.) But it is easy to see that this definition is equivalent to the one given above.

The following result, which is Theorem 1.3 in Roberts (1979) will be used in section 2.5 about semiorders.

Lemma 2.3.1: The binary relation (A,ρ) is a strict weak order iff (A,ρ) is asymmetric and transitive and (A,σ) is an equivalence relation, where σ is defined by

$$aσb \quad iff \quad ¬aρb \text{ and } ¬bρa.$$

(For proof see Roberts, 1979, p. 33.)

As can be seen from the above list an order which is classified as strict is asymmetric. A non-strict order is normally symmetric. Note the following connections between strict and non-strict orders.

(1) Suppose $\langle A,\rho \rangle$ is a weak order. Define π on A by

$$a\pi b \quad \text{iff} \quad (a\rho b \;\&\; \neg b\rho a).$$

Then $\langle A,\pi \rangle$ is a strict weak order. (See Roberts, 1979, p. 36.)

(2) Suppose $\langle A,\pi \rangle$ is a strict weak order and σ is defined on A by

$$a\sigma b \quad \text{iff} \quad (\neg a\pi b \;\&\; \neg b\pi a).$$

Then $\langle A,\sigma \rangle$ is an equivalence relation. And further, $\langle A,\rho \rangle$ where ρ is defined on A by

$$a\rho b \quad \text{iff} \quad (a\pi b \text{ or } a\sigma b)$$

is a weak order. (Use 2.3.1.)

(3) Suppose (A,π) is an irreflexive binary relation. Define ρ on A by

$$a\rho b \quad \text{iff} \quad (a\pi b \text{ or } a=b).$$

Then (A,π) is a strict simple order if and only if (A,ρ) is a simple order. (See Roberts, 1979, p. 32.)

(4) Suppose $\langle A,\pi \rangle$ is an irreflexive binary relation. Define ρ on A by

$$a\rho b \quad \text{iff} \quad (a\pi b \text{ or } a=b).$$

Then $\langle A,\pi \rangle$ is a strict partial order iff $\langle A,\rho \rangle$ is a partial order. (See Roberts, 1979, p. 38.)

For a nice presentation of relations and orders see Roberts (1979) chapter 1.

2.4 Two lemmas on weak orders

In this section we state two simple lemmas which will be used in this and later chapters. The first at least states something well known.

Lemma 2.4.1: If $\langle A,\rho \rangle$ is a weak order and π and σ are defined by

$$a\pi b \quad \text{iff} \quad (a\rho b \;\&\; \neg b\rho a).$$
$$a\sigma b \quad \text{iff} \quad (\neg a\pi b \;\&\; \neg b\pi a)$$

then it holds that

(1) $a\sigma b$ iff $a\rho b \;\&\; b\rho a$
(2) $a\pi b$ iff $a\rho b \;\&\; \neg a\sigma b$
(3) $a\rho b$ iff $a\pi b$ or $a\sigma b$.

The straightforward proof is omitted.

Lemma 2.4.2: Suppose that $\langle A,\rho \rangle$ and $\langle A,\rho' \rangle$ are weak orders. Let π and σ be defined from ρ as in the preceding lemma; define π' and σ' analogously from ρ'.

(i) If $\rho \subseteq \rho'$, then $\pi \supseteq \pi'$ and $\sigma \subseteq \sigma'$. And if $\rho \subset \rho'$, then $\pi \supset \pi'$ and $\sigma \subset \sigma'$.

(ii) If $\pi \subseteq \pi'$, then $\rho \supseteq \rho'$ and $\sigma \supseteq \sigma'$. And if $\pi \subset \pi'$, then $\rho \supset \rho'$ and $\sigma \supset \sigma'$.

Proof: We first prove (i).

(1) Suppose $\rho \subseteq \rho'$. Suppose further that $a\pi'b$. Hence, $a\rho'b$ & $\neg b\rho'a$. If $b\rho a$ then $b\rho'a$ according to the assumption that $\rho \subseteq \rho'$, but this implies a contradiction. Therefore $\neg b\rho a$. Since ρ is a weak order it is connected so $\neg b\rho a$ implies $a\rho b$. Hence, $a\rho b$ & $\neg b\rho a$, i.e. $a\pi b$. We have thus proved that $a\pi'b$ implies $a\pi b$, i.e. $\pi' \subseteq \pi$. Hence, $\rho \subseteq \rho'$ implies $\pi \supseteq \pi'$.

Suppose now that $\rho \subset \rho'$. Let a and b be such that $\neg a\rho b$ and $a\rho'b$. Since ρ is connected $\neg a\rho b$ implies $b\rho a$. Hence, $b\pi a$. $a\rho'b$ implies $\neg b\pi'a$. We have thus proved that $\neg(\pi \subseteq \pi')$. Since $\rho \subseteq \rho'$ implies $\pi \supseteq \pi'$ we get $\pi \supset \pi'$. Therefore $\rho \subset \rho'$ implies $\pi \supset \pi'$.

(2) Suppose $\rho \subseteq \rho'$. Suppose further that $a\sigma b$. Then $a\rho b$ and $b\rho a$. From the assumption that $\rho \subseteq \rho'$ follows $a\rho'b$ and $b\rho'a$, which implies $a\sigma'b$. We have thus proved that $\sigma \subseteq \sigma'$. Hence, $\rho \subseteq \rho'$ implies $\sigma \subseteq \sigma'$.

Suppose now that $\rho \subset \rho'$. Let a and b be such that $\neg a\rho b$ and $a\rho'b$. Hence $b\rho a$, which implies $b\rho'a$. We have therefore $\neg a\sigma b$ and $a\sigma'b$, which shows that $\neg(\sigma' \subseteq \sigma)$. Since $\rho \subseteq \rho'$ implies $\sigma \subseteq \sigma'$ it follows that $\rho \subset \rho'$ implies $\sigma \subset \sigma'$.

We now prove (ii).

(1) Suppose $\pi \subseteq \pi'$. Suppose further that $a\sigma'b$. Then $\neg a\pi'b$ and $\neg b\pi'a$. Hence, $\neg a\pi b$ and $\neg b\pi a$, i.e. $a\sigma b$. We have thus proved that $\pi \subseteq \pi'$ implies $\sigma' \subseteq \sigma$.

Suppose now that $\pi \subset \pi'$. Let a and b be such that $a\pi'b$ but $\neg a\pi b$. $a\pi'b$ implies $\neg b\pi'a$ since π' is asymmetric, and $\neg b\pi'a$ implies $\neg b\pi a$ according to the assumption. Hence, $\neg a\pi b$ and $\neg b\pi a$, i.e. $a\sigma b$. But $\neg a\sigma'b$ since $a\pi'b$. Therefore, $\neg(\sigma' \supseteq \sigma)$. Since $\pi \subset \pi'$ implies $\sigma' \subseteq \sigma$, it holds that $\pi \subset \pi'$ implies $\sigma' \subset \sigma$.

(2) Suppose $\pi \subseteq \pi'$. Suppose further that $a\rho'b$. Then $a\pi'b$ or $a\sigma'b$. Consider first the case that $a\sigma'b$. Then $a\rho b$, since $\pi \subseteq \pi'$ implies $\sigma' \subseteq \sigma$ and $a\sigma b$ is equivalent to $a\rho b$ and $b\rho a$. Consider now the case that $a\pi'b$. Then $\neg b\pi'a$, since π is asymmetric, which implies $\neg b\pi a$ (since $\pi \subseteq \pi'$). There are now two possibilities: $a\pi b$ or $\neg a\pi b$. Since $\neg b\pi a$, both possibilities imply $a\rho b$. We have thus proved that $\pi \subseteq \pi'$ implies $\rho' \subseteq \rho$.

Suppose now that $\pi \subset \pi'$. Let a and b be such that $a\pi'b$ but $\neg a\pi b$. From aπ'b follows ¬bρ'a. If bπa then bπ'a, which implies ¬aπ'b since π' is asymmetric. Therefore ¬bπa, ¬aπb and ¬bπa imply bρa. Hence ¬(ρ⊆ρ'). Since π ⊂ π' implies ρ'⊆ ρ, it follows that ρ'⊂ ρ. We have thus proved that π ⊂ π' implies ρ'⊂ ρ. ♦

Note that σ ⊆ σ' does not imply either ρ ⊆ ρ' or ρ'⊆ ρ. To see this suppose that ρ and ρ' are as in the figure below.

$$\underline{ρ} \qquad\qquad \underline{ρ'}$$
$$a \qquad\qquad b$$
$$b \qquad\qquad a$$

Hence σ= σ'= ∅ but ρ={⟨a,b⟩,⟨a,a⟩,⟨b,b⟩} and ρ'={⟨b,a⟩,⟨a,a⟩,⟨b,b⟩}. Therefore σ ⊆ σ' but ¬(ρ⊆ρ') and ¬(ρ'⊆ ρ).

2.5. Semiorders

A binary relation ⟨A,ζ⟩ is called a *semiorder* if the following axioms are satisfied for all a,b,c,d in A.

(i) ¬aζa

(ii) aζb & cζd ⇒ (aζd or cζb)

(iii) aζb & bζc ⇒ (aζd or dζc)

The concept of a semiorder was introduced by Luce (1956). The definition above is the one given by Scott and Suppes (1958).

Every semiorder is a strict partial order, i.e. asymmetric and transitive. To see that transitivity holds, take d=c in axiom (iii). We then get

$$aζb \ \& \ bζc \ ⇒ \ aζc \ or \ cζc .$$

From axiom (i) it follows that ¬cζc, which implies that ζ is transitive. Asymmetry follows from transitivity and irreflexiveness; aζb and bζa implies aζa, which contradicts axiom (i).

It is often maintained that relations like *definitely louder than* and *definitely higher than* used of sounds, *definitely brighter than* used of hues of colour, *noticeably greater than* used of weights and so on are semiorders. (See Scott and Suppes, 1958, p. 117.)

Let ⟨A,ζ⟩ be a semiorder. Define the binary relation ⟨A,δ⟩ by

$$aδb \ iff \ ¬aζb \ \& \ ¬bζa. \qquad\qquad (2.5.1)$$

δ can be regarded as an indifference relation. Note that δ is not necessarily an equivalence relation since δ may not be transitive. As a matter of fact the following holds: If $\langle A,\zeta\rangle$ is a semiorder and δ is defined by Equation (2.5.1), then $\langle A,\zeta\rangle$ is a strict weak order iff $\langle A,\delta\rangle$ is an equivalence relation. This follows immediately from 2.3.1 since a semiorder is always asymmetric and transitive.

If $\langle A,\zeta\rangle$ is a semiorder, then δ defined as above is thus not necessarily an equivalence relation. But the binary relation ε defined from δ by

$$a\varepsilon b \text{ iff } \forall c\in A: a\delta c \Leftrightarrow b\delta c$$

is an equivalence relation.

Given a semiorder $\langle A,\zeta\rangle$ we can define a weak order $\langle A,\rho\rangle$ in the following way:

$$a\rho b \text{ iff } (\forall c\in A)[(c\zeta a \Rightarrow c\zeta b) \ \& \ (b\zeta c \Rightarrow a\zeta c)]$$

ρ is called the *weak order associated with* ζ. This concept is originally due to Luce (1956) but we follow here the presentation in Scott and Suppes (1958), which also is represented in Roberts (1979) chapter 6.

Since $\langle A,\rho\rangle$ is a weak order we can define a strict weak order $\langle A,\pi\rangle$ and an equivalence relation $\langle A,\sigma\rangle$ in the following way:

$$a\pi b \text{ iff } a\rho b \text{ and } \neg b\rho a$$
$$a\sigma b \text{ iff } a\rho b \text{ and } b\rho a.$$

We call π the *strict weak order associated with* ζ and σ the *equivalence relation associated with* ζ . Note that

$$a\sigma b \text{ iff } (\forall c\in A)[(c\zeta a \Leftrightarrow c\zeta b) \ \& \ (b\zeta c \Leftrightarrow a\zeta c)]$$
$$a\pi b \text{ iff } (\forall c\in A)[(c\zeta a \Rightarrow c\zeta b) \ \& \ (b\zeta c \Rightarrow a\zeta c)] \ \& \ (\exists c\in A)[(c\zeta b \ \& \ \neg c\zeta a) \vee$$
$$\vee (a\zeta c \ \& \ \& \ \neg b\zeta c)].$$

It is a well-known fact, and easy to see, that σ and ε (as defined above) coincide. We thus have two alternative ways of defining the equivalence relation associated to a semiorder.

The following theorem will be used in section 3.3.

Theorem 2.5.2: Suppose $\langle A,\zeta\rangle$ is a semiorder and $B\subseteq A$ and $\eta=\zeta/B$. Then

(i) $\langle B,\eta\rangle$ is a semiorder

(ii) for the weak orders ρ associated with ζ and υ associated with η it holds that

$$\upsilon \supseteq \rho/B.$$

(iii) for the equivalence relations σ associated with ζ and τ associated with η it holds that

$$\tau \supseteq \sigma/B$$

(iv) for the strict weak orders π associated with ζ and ω associated with η it holds that

$$\omega \subseteq \pi .$$

Proof: (i) It is obvious from the axioms for a semiorder that $\langle B,\eta \rangle$ is a semiorder. (ii) Suppose that $\langle a,b \rangle \in \rho/B$, i.e. $a\rho b$ and $a,b \in B$. This implies, according to the definition of $a\rho b$, that

$$(\forall c \in A)\ [(c\zeta a \Rightarrow c\zeta b)\ \&\ (b\zeta c \Rightarrow a\zeta c)] .$$

Therefore, since $\eta = \zeta/B$,

$$(\forall c \in B)\ [(c\eta a \Rightarrow c\eta b)\ \&\ (b\eta c \Rightarrow a\eta c)]$$

i.e. $a\upsilon b$. We have thus shown that

$$\rho/B \subseteq \upsilon .$$

We prove (iii) and (iv) from (ii) using 2.4.2. $\langle B,\rho/B \rangle$ and $\langle B,\upsilon \rangle$ are weak orders and $\rho/B \subseteq \upsilon$. Since π is the strict weak order and σ the equivalence relation corresponding to ρ, it follows that π/B is the strict weak order and σ/B the equivalence relation corresponding to ρ/B. Then it follows from 2.4.2 that $\pi/B \supseteq \omega$ and $\sigma/B \subseteq \tau$. ♦

A more extensive treatment of semiorders can be found in for example Luce (1956), Scott and Suppes (1958), Suppes and Zinnes (1963) p. 29f. and Roberts (1979) section 6.1.

2.6 Correspondences

A *correspondence* from a set X to (or into) a set Y is a subset of the Cartesian product $X \times Y$, i.e. any binary relation from X to Y. $X \times Y$ is the *universal correspondence* from X to Y. $\varnothing$ is the *empty* correspondence from X to Y. If γ is a correspondence from X to Y and A is any subset of X, we define

$$\gamma[A] = \{y \in Y \mid \langle x,y \rangle \in \gamma \text{ for some } x \in A\} .$$

As a generalization of this, we define $\gamma[\rho]$, where ρ is an n-ary relation in A, as follows:

$$\{\langle a_1,...,a_n \rangle \in A^n \mid \exists x_1,...,x_n \in A\colon x_1\gamma a_1\ \&\ ...\ \&\ x_n\gamma a_n\ \&\ \rho(x_1,...,x_n)\}.$$

$\langle x,y \rangle \in \gamma$ will often be written $x\gamma y$. With the converse γ^{-1} of a correspondence γ is meant the set $\{\langle x,y \rangle \mid \langle y,x \rangle \in \gamma\}$.

Suppose γ is a correspondence from X to Y. The set $\{x \in X \mid \exists y \in Y\colon x\gamma y\}$ is called the *domain* of the correspondence γ. And the set $\{y \in Y \mid \exists x \in X\colon x\gamma y\}$ is

called the *range* of γ. Note that if A is the domain of the correspondence γ then
γ[A] is the range of γ. And if B is the range of the correspondence γ then γ⁻¹[B] is
the domain of γ. If A is the domain of γ then γ is a correspondence *on* A to Y. If
B is the range of γ then γ is a correspondence from X *onto* B. If A is the domain
of γ and B is the range of γ, then γ is a correspondence *on* A *onto* B.

If γ is a correspondence with domain X, and to any $x \in X$ there corresponds one
and only one $y \in Y$ (i.e. γ[{x}] consists of exactly one element for any x in X), then
γ is called a *univalent* correspondence. If γ and γ⁻¹ are both univalent correspondences, then γ is called a *one-to-one correspondence*. A univalent correspondence is thus a function. And a one-to-one correspondence is a function such
that its inverse is a function. (γ is a function iff whenever $\langle x,y \rangle \in \gamma$ and $\langle x,z \rangle \in \gamma$
then y=z.)

For any correspondence γ and δ we define the relative product of γ and δ,
written $\gamma \mid \delta$, by

$$\gamma \mid \delta = \{\langle x,y \rangle \mid \exists z : x\gamma z \ \& \ z\delta y\}$$

Note that if γ is a correspondence from X to Y and δ a correspondence from A
to B, then $\gamma \mid \delta$ is a correspondence from X to B. And $\gamma \mid \delta = \emptyset$ unless $Y \cap A \neq \emptyset$.

Since a correspondence is just a binary relation it may seem superfluous to
introduce this notion. But when we are interested in the "mapping character" of a
binary relation, especially in situations where a binary relation turns out to be a
function, then the correspondence-terminology is appropriate.

Some simple results on correspondences are collected in the following lemma.
They will be used in this and later chapters.

Lemma 2.6.1: Suppose $\gamma, \delta \subseteq X \times Y$ and $A, B \subseteq X$. Then
(1) γ[A∩B] $\subseteq$ γ[A] ∩ γ[B].
(2) γ[A∪B] = γ[A] ∪ γ[B].
(3) $(\gamma \mid \delta)[A] = \delta[\gamma[A]]$
(4) $\gamma \supseteq \delta \Rightarrow$ γ[A] $\supseteq$ δ[A].
(4) γ[A] = $\bigcup_{x \in A}$ γ[{x}].

The following two lemmas will be used in section 2.7. Proofs are omitted.

Lemma 2.6.2: Suppose ρ is a n-ary relation on A and γ a correspondence from A
to A. Then

$$\gamma[\rho/\gamma^{-1}[A]]=\gamma[\rho].$$

Lemma 2.6.3: Suppose ρ is a relation on A and γ a function from A into A. Then $\gamma[\gamma^{-1}[\rho]]\subseteq\rho$.

2.7 Invariance

In this section we shall apply the notion of invariance to relations (relations in the sense of sets of ordered n-tuples). This means that that which remains invariant— Margenau's "invariant property" (see section 1.6)—is a relation. The class of "changes" under which invariance of the relation is said to hold—Margenau's "transformations"—we wish to construe as widely as possible. We shall consider whether it is reasonable to talk about the invariance of a relation on a set A under a correspondence from A to A. Usually, the invariance of a relation on a set A seems to be with respect to a function from A into A, or more specifically, a bijection of A. We shall return to the question of why it is interesting to generalize to correspondences later in this chapter.

In section 1.6 two definitions of invariance found in Weyl (1949) was quoted. The second one has the following wording.

> A ternary relation R(xyz) between points is *invariant* with respect to a given transformation $\sigma:p\rightarrow p'$ and its inverse $p'\rightarrow p$ if R(abc) always implies R(a'b'c') and vice versa.

Weyl defines the invariance of a relation with respect to a transformation and its inverse. With Weyl's definition in mind the following definition of invariance relative to a bijection would be tempting. The n-ary relation ρ on A is *invariant with respect to* the bijection φ on A if $\rho(a_1,...a_n)$ implies $\rho(\varphi(a_1),...,\varphi(a_n))$. According to this definition ρ is invariant with respect to the bijection φ on A and its inverse φ^{-1} iff

$$\rho(a_1,...,a_n) \Leftrightarrow \rho(\varphi(a_1),...,\varphi(a_n)).$$

The usual way of defining invariance relative to a bijection seems, however, not to be this, but rather the following: The n-ary relation ρ on A is *invariant under* the bijection φ on A if for all $a_1,...a_n\in A$

$$\rho(a_1,...,a_n) \Leftrightarrow \rho(\varphi(a_1),...,\varphi(a_n)).$$

Note that we say here that ρ is invariant *under* φ to distinguish this definition from Weyl's.

The advantage with the last definition can be seen from 2.7.1. First a definition. Let γ be a correspondence from A to A. We define $\gamma[\rho]$ in the following way:

$\gamma[\rho] = \{\langle a_1,...,a_n \rangle \in A^n \mid \exists x_1,...,x_n \in A: x_1 \gamma a_1 \ \& \ ... \ \& \ x_n \gamma a_n \ \& \ \rho(x_1,...,x_n)\}.$

Note that a function on A into A is a correspondence from A to A. If φ is a function on A into A, then

$\varphi[\rho] = \{\langle a_1,...,a_n \rangle \in A^n \mid \exists x_1,...,x_n \in A: \varphi(x_1)=a_1 \ \&...\& \ \varphi(x_n)=a_n \ \& \ \rho(x_1,...,x_n)\}.$

Lemma 2.7.1: If ρ is a relation on A and φ is a bijection on A, then ρ is invariant under φ iff $\varphi[\rho]=\rho$.

Proof: (I) Suppose that ρ is invariant under φ.

(1) Suppose further that $\langle a_1,...,a_n \rangle \in \varphi[\rho]$. Then there is $\langle x_1,...,x_n \rangle \in A$ such that $\varphi(x_1)=a_1,...,\varphi(x_n)=a_n$ and $\rho(x_1,...,x_n)$. Since ρ is invariant under φ it follows that $\rho(\varphi(x_1),...,\varphi(x_n))$. Hence $\rho(a_1,...,a_n)$. We have thus proved that $\varphi[\rho] \subseteq \rho$.

(2) Suppose now that $\rho(a_1,...,a_n)$. Since φ is a bijection it follows that $\rho(\varphi(\varphi^{-1})(a_1),...,\varphi(\varphi^{-1})(a_n))$. From the assumption that ρ is invariant under φ it follows that $\rho(\varphi^{-1}(a_1),...,\varphi^{-1}(a_n))$. Then $\langle a_1,...,a_n \rangle \in \varphi[\rho]$. Hence, $\rho \subseteq \varphi[\rho]$. We have thus proved that if ρ is invariant under φ then $\varphi[\rho]=\rho$.

(II) Suppose that $\varphi[\rho]=\rho$.

(1) Suppose further that $\rho(a_1,...,a_n)$. Hence $\langle \varphi(a_1),...,\varphi(a_n) \rangle \in \varphi[\rho]$. Since $\varphi[\rho]=\rho$ it follows that $\langle \varphi(a_1),...,\varphi(a_n) \rangle \in \rho$.

(2) Suppose that $\rho(\varphi(a_1),...,\varphi(a_n))$. Then there exist $x_1,...,x_n \in A$ such that $\varphi(x_1)=\varphi(a_1),...,\varphi(x_n)=\varphi(a_n)$ and $\rho(x_1,...,x_n)$. Since φ is a bijection it follows that $a_1=x_1,...,a_n=x_n$. Hence, $\rho(a_1,...,a_n)$. We have thus proved that if $\varphi[\rho]=\rho$ then $\rho(a_1,...,a_n)$ is equivalent to $\rho(\varphi(a_1),...,\varphi(a_n))$. ◆

That ρ is invariant under φ means informally that ρ is not changed by the transformation φ. The question is now how this should be interpreted. One suggestion is $\varphi[\rho]=\rho$. If we accept this the preceding lemma gives us a reason for choosing the definition adopted there, i.e. *invariance under*. It is easy to see from the proof of the preceding lemma that ρ is invariant *with respect to* φ iff $\varphi[\rho] \subseteq \rho$.

It is natural to ask if it is essential that the invariance of a relation is relative to a *bijection*. In Pfanzagl (1971) p. 36 there is a definition of invariance of a relation on a set A relative to a function on a subset of A into A. Transformed into the notation used here the definition runs as follows. If ρ is an n-ary relation on A and φ is a function on a subset B of A into A, then ρ is φ-*invariant* if for all for all $a_1,...a_n \in B$

$$\rho(a_1,...,a_n) \Leftrightarrow \rho(\varphi(a_1),...,\varphi(a_n)).$$

This is of course a generalization of invariance under bijections. But note that lemma 2.7.1 cannot be generalized to hold for φ-invariance, i.e. it does not hold that ρ is φ-invariant iff $\varphi[\rho]=\rho/\varphi(B)$. For it does not follow that $\varphi[\rho]=\rho/\varphi(B)$ implies

$$\rho(\varphi(a_1),...,\varphi(a_n)) \Rightarrow \rho(a_1,...,a_n).$$

To give a concrete example, let $A=\{a,b,c\}$ and $\rho=\{\langle a,b\rangle\}$ and $\varphi=\{\langle a,a\rangle,\langle b,b\rangle,\langle c,a\rangle\}$. Then φ is a function on A into A and $\varphi[\rho]=\{\langle a,b\rangle\}=\rho/\varphi[A]$. But $\rho(\varphi(c),\varphi(b))$ and $\neg\rho(c,b)$.

However, Pfanzagl point out that the definition of φ-invariance is equivalent to $\rho/B=\varphi^{-1}[\rho]$. Pfanzagl's definition of invariance therefore means that ρ is unchanged by φ^{-1} rather than by φ. There is something here which ought to be cleared up, especially if we want to generalize the notion of invariance so that a relation on a set A could be invariant under a correspondence of A. With this aim in mind we prove some simple lemmas and theorems.

Lemma 2.7.2: Suppose ρ is a n-ary relation on A and γ a correspondence from A to A. Then the following two conditions are equivalent.
(1) for all $a_1,...,a_n,b_1,...,b_n \in A$ such that $a_1\gamma b_1,...,a_n\gamma b_n$:

$$\rho(a_1,...,a_n) \Rightarrow \rho(b_1,...,b_n)$$

and

(2) $\gamma[\rho]\subseteq\rho/\gamma[A]$.

Proof: (I) Suppose condition (1). Suppose further that $\langle a_1,...,a_n\rangle\in\gamma[\rho]$. Then there exist $x_1,...,x_n\in A$ such that $x_1\gamma a_1, ...,x_n\gamma a_n$ and $\rho(x_1,...,x_n)$. From this it follows, according to the assumption, that $\rho(a_1,...,a_n)$. Since $a_1,...,a_n\in\gamma[A]$ it follows that $\langle a_1,...,a_n\rangle\in\rho/\gamma[A]$.

(II) Suppose condition (2). Suppose further that $\rho(a_1,...,a_n)$ and $a_1\gamma b_1,...,a_n\gamma b_n$. Then $\langle b_1,...,b_n\rangle\in\gamma[\rho]$. From the assumption it then follows that $\langle b_1,...,b_n\rangle\in\rho/\gamma[A]$. Hence $\rho(b_1,...,b_n)$. ♦

Lemma 2.7.3: Suppose ρ is a n-ary relation on A and γ a correspondence from A to A. Then the following two conditions are equivalent.
(1) for all $a_1,...,a_n,b_1,...,b_n \in A$ and $b_1\gamma a_1,...,b_n\gamma a_n$,

$$\rho(a_1,...,a_n) \Rightarrow \rho(b_1,...,b_n)$$

and

(2) $\gamma^{-1}[\rho] \subseteq \rho/\gamma^{-1}[A]$.

Proof: Since γ is a correspondence from A to A it follows that γ^{-1} is a correspondence from A to A. Condition (1) is equivalent to

(1') for all $a_1,...,a_n,b_1,...,b_n \in A$ such that $a_1\gamma^{-1}b_1,...,a_n\gamma^{-1}b_n$,

$$\rho(a_1,...,a_n) \Rightarrow \rho(b_1,...,b_n)$$

which according to 2.7.2 is equivalent to

(2) $\gamma^{-1}[\rho] \subseteq \rho/\gamma^{-1}[A]$. ◆

Theorem 2.7.4: Suppose ρ is a n-ary relation on A and γ a correspondence from A to A. Then the following two conditions are equivalent:

(1) for all $a_1,...,a_n,b_1,...,b_n \in A$ such that $a_1\gamma b_1,,...,a_n\gamma b_n$,

$$\rho(a_1,...,a_n) \Leftrightarrow \rho(b_1,...,b_n)$$

and

(2) $\gamma[\rho] \subseteq \rho/\gamma[A]$ and $\gamma^{-1}[\rho] \subseteq \rho/\gamma^{-1}[A]$.

Proof: Apply 2.7.2 and 2.7.3. ◆

Lemma 2.7.5: Suppose ρ is a relation on A and γ a correspondence from A to A. Then $\gamma^{-1}[\rho] \subseteq \rho/\gamma^{-1}[A]$ implies $\rho/\gamma[A] \subseteq \gamma[\rho]$.

Proof: Suppose $\gamma^{-1}[\rho] \subseteq \rho/\gamma^{-1}[A]$. Suppose further that $\langle a_1,...,a_n \rangle \in \rho/\gamma[A]$. Then $\rho(a_1,...,a_n)$ and there are $b_1,...,b_n \in A$ such that $b_1\gamma a_1,...,b_n\gamma a_n$. Hence, $\langle b_1,...,b_n \rangle \in \gamma^{-1}[\rho]$ and according to the assumption $\rho(b_1,...,b_n)$. Since $b_1\gamma a_1,...,b_n\gamma a_n$ it follows that $\langle a_1,...,a_n \rangle \in \gamma[\rho]$. ◆

The converse of 2.7.5 does not hold, as the following example shows. Let $A=\{a,b,c\}$ and $\rho=\{\langle a,b \rangle\}$ and $\gamma=\{\langle a,a \rangle,\langle b,b \rangle,\langle c,a \rangle\}$. Then $\rho/\gamma[A]=\gamma[\rho]$ but $\gamma^{-1}[\rho] \supset \rho/\gamma^{-1}[A]$, since $\gamma^{-1}[\rho]=\{\langle a,b \rangle,\langle c,b \rangle\}$ and $\rho/\gamma^{-1}[A]=\rho$.

Lemma 2.7.6: Suppose ρ is a relation on A and γ a correspondence from A to A. Then $\gamma[\rho] \subseteq \rho/\gamma[A]$ implies $\rho/\gamma^{-1}[A] \subseteq \gamma^{-1}[\rho]$.

Proof Since γ^{-1} is a correspondence from A to A and $(\gamma^{-1})^{-1}=\gamma$ then, according to 2.7.5, $\gamma[\rho] \subseteq \rho/\gamma[A]$ implies $\rho/\gamma^{-1}[A] \subseteq \gamma^{-1}[\rho]$. ◆

The converse of 2.7.6 does not hold, as the following example shows. Let $A=\{a,b,c\}$ and $\rho=\{\langle a,b \rangle\}$ and $\gamma=\{\langle a,a \rangle,\langle b,b \rangle,\langle a,c \rangle\}$. Then $\rho/\gamma^{-1}[A]=\gamma^{-1}[\rho]$ but $\gamma[\rho] \supset \rho/\gamma[A]$, since $\gamma[\rho]=\{\langle a,b \rangle,\langle c,b \rangle\}$ and $\rho/\gamma[A]=\rho$.

Theorem 2.7.7: Suppose ρ is a relation on A and γ a correspondence from A to A. Then the condition

(1) for all $a_1,...,a_n,b_1,...,b_n \in A$ such that $a_1\gamma b_1,,...,a_n\gamma b_n$,

$$\rho(a_1,...,a_n) \Leftrightarrow \rho(b_1,...,b_n)$$

is equivalent to

(2) $\gamma[\rho]=\rho/\gamma[A]$ and $\gamma^{-1}[\rho]=\rho/\gamma^{-1}[A]$.

Proof: Follows from 2.7.4, 2.7.5 and 2.7.6. ◆

Note that neither of $\gamma[\rho]=\rho/\gamma[A]$ and $\gamma^{-1}[\rho]=\rho/\gamma^{-1}[A]$ taken by itself implies (1).

Lemma 2.7.8: Suppose ρ is a relation on A and γ a function from A into A. Then $\rho/\gamma^{-1}[A] \subseteq \gamma^{-1}[\rho]$ implies $\gamma[\rho] \subseteq \rho/\gamma[A]$.

Proof: Suppose $\rho/\gamma^{-1}[A] \subseteq \gamma^{-1}[\rho]$. Then $\gamma[\rho/\gamma^{-1}[A]] \subseteq \gamma[\gamma^{-1}[\rho]]$. According to 2.6.2 $\gamma[\rho/\gamma^{-1}[A]]=\gamma[\rho]$. Since γ is a function it follows from 2.6.3 that $\gamma[\gamma^{-1}[\rho]] \subseteq \rho$. Hence, $\gamma[\rho] \subseteq \rho$ and since $\gamma[\rho]=\gamma[\rho]/\gamma[A] \subseteq \rho/\gamma[A]$ it follows that $\gamma[\rho] \subseteq \rho/\gamma[A]$. ◆

Lemma 2.7.9: Suppose ρ is a relation on A and γ^{-1} a function from A into A. Then $\rho/\gamma[A] \subseteq \gamma[\rho]$ implies $\gamma^{-1}[\rho] \subseteq \rho/\gamma^{-1}[A]$.

Proof: According to 2.7.8, $\rho/(\gamma^{-1})^{-1}[A] \subseteq (\gamma^{-1})^{-1}[\rho]$ implies $\gamma^{-1}[\rho]=\rho/\gamma^{-1}[A]$ and since $(\gamma^{-1})^{-1}=\gamma$ the lemma follows. ◆

Lemma 2.7.10: Suppose ρ is a relation on A.

(i) If γ is a function from A into A, then $\rho/\gamma^{-1}[A]=\gamma^{-1}[\rho]$ implies $\gamma[\rho]=\rho/\gamma[A]$.

(ii) If γ^{-1} is a function from A into A, then $\rho/\gamma[A]=\gamma[\rho]$ implies $\gamma^{-1}[\rho]=\rho/\gamma^{-1}[A]$.

(iii) If γ is a one-to-one-function from A into A, then $\gamma[\rho]=\rho/\gamma[A]$ is equivalent to $\gamma^{-1}[\rho]=\rho/\gamma^{-1}[A]$.

(iv) If γ is a bijection of A, then $\gamma[\rho]=\rho/\gamma$ is equivalent to $\gamma^{-1}[\rho]=\rho/\gamma^{-1}$.

Proof: For the proof of (i) use 2.7.5 and 2.7.8. For the proof of (ii) use 2.7.6 and 2.7.9. For the proof of (iii) use (i) and (ii). For the proof of (iv) note that if γ is a bijection on A then $\gamma[A]=A=\gamma^{-1}[A]$. ◆

Theorem 2.7.11: Suppose ρ is a relation on A and γ a function from A into A. Then the following two conditions are equivalent:

(1) for all $a_1,...,a_n \in \gamma^{-1}[A]$,

$$\rho(a_1,...,a_n) \Leftrightarrow \rho(\gamma(b_1),...,\gamma(b_n))$$

and

(2) $\gamma^{-1}[\rho]=\rho/\gamma^{-1}[A]$.

Proof: Use 2.7.7 and 2.7.10(i). ♦

Theorem 2.7.12: Suppose ρ is a relation on A and γ^{-1} a function from A into A. Then the following two conditions are equivalent:

(1) for all $a_1,...,a_n,b_1,...,b_n \in A$ such that $a_1\gamma b_1,,...,a_n\gamma b_n$:

$$\rho(a_1,...,a_n) \Leftrightarrow \rho(b_1,...,b_n)$$

and

(2) $\gamma[\rho]=\rho/\gamma[A]$.

Proof: Use 2.7.7 and 2.7.10(ii). ♦

Theorem 2.7.13: Suppose ρ is a relation on A and γ a bijection on A. Then the following three conditions are equivalent:

(1) for all $a_1,...,a_n \in A$,

$$\rho(a_1,...,a_n) \Leftrightarrow \rho(\gamma(a_1),...,\gamma(a_n))$$

(2) $\gamma^{-1}[\rho]=\rho$

(3) $\gamma[\rho]=\rho$.

Proof: Use 2.7.10(iv), 2.7.11 and 2.7.12. ♦

Let us now turn to the problem of how invariance of a relation under a correspondence could reasonably be defined. The following question is important. What could it mean for an n-ary relation ρ on A to be constant under the changes carried out by a correspondence γ on A? $\gamma[\rho]=\rho/\gamma[A]$ seems to be a necessary condition. But is it sufficient? I would say no. It seems reasonable to also demand that if $\neg\rho(a_1,...,a_n)$ then the changes accomplished by γ should not change the fact that γ does not apply, i.e. if $\neg\rho(a_1,...,a_n)$ and $a_1\gamma b_1,,...,a_n\gamma b_n$ then $\neg\rho(b_1,...,b_n)$. This implies according to 2.7.3 that $\gamma^{-1}[\rho]\subseteq\rho/\gamma^{-1}[A]$, which together with $\gamma[\rho]=\rho/\gamma[A]$ implies that $\gamma^{-1}[\rho]=\rho/\gamma^{-1}[A]$ according to 2.7.6. I therefore suggest that an n-ary relation ρ on A is constant under the changes carried out by a correspondence γ on A if and only if $\gamma[\rho]=\rho/\gamma[A]$ and $\gamma^{-1}[\rho]=\rho/\gamma^{-1}[A]$. We can now use 2.7.7 and arrive at the following definition of invariance of a relation under a correspondence.

An n-ary relation ρ on A is *invariant under* the correspondence γ from A to A if for all $a_1,...,a_n,b_1,...,b_n \in A$ such that $a_1\gamma b_1,,...,a_n\gamma b_n$,

$$\rho(a_1,...,a_n) \Leftrightarrow \rho(b_1,...,b_n).$$

One of the reasons for studying invariance under correspondences is clear from the first quotation from Weyl (1949) in section 1.6, which contains a definition of invariance. For convenience we repeat it here.

> R(xy) is invariant with respect to the equivalence ≈, if R(ab) always entails R(a'b'), provided a'≈a and b'≈b.

Weyl's definition can in the terminology used here be formulated as follows:
An n-ary relation ρ on A is invariant with respect to the equivalence relation γ on A if for all $a_1,...,a_n,b_1,...,b_n \in$ A such that $b_1\gamma a_1,...,b_n\gamma a_n$,

$$\rho(a_1,...,a_n) \Rightarrow \rho(b_1,...,b_n).$$

Since an equivalence relation on A is a correspondence from A to A we can apply 2.7.3, so Weyl's definition is equivalent to $\gamma^{-1}[\rho]\subseteq\rho/\gamma^{-1}[A]$. As an equivalence relation, γ is reflexive, symmetric and transitive. The symmetry of γ implies that $\gamma=\gamma^{-1}$ and the reflexivity of γ implies that $\gamma[A] = \gamma^{-1}[A]=A$. Hence, $\gamma[\rho]\subseteq\rho/\gamma[A]$ implies $\gamma[\rho]\subseteq\rho$ and $\gamma^{-1}[\rho]\subseteq\rho$. From the last fact it follows according to 2.7.5 that $\rho\subseteq\gamma[\rho]$. Weyl's definition thus implies that $\rho=\gamma[\rho]$ and, since $\gamma=\gamma^{-1}$, $\rho=\gamma^{-1}[\rho]$. This shows, according to 2.7.7, that Weyl's definition of invariance with respect to an equivalence relation is just a special case of invariance under a correspondence.

Note that if a relation ρ is invariant under an equivalence relation γ, then γ is a congruence relation for ρ. This means that we can use *invariance under a correspondence* to characterize congruence relations. We shall develop this line of thought in section 2.10.

The notion of invariance can be straightforwardly extended so that it holds with respect to a *set* of correspondences. Let ρ be a relation on A and Γ a set of correspondences from A to A. We say that ρ is *invariant under* Γ iff ρ is invariant under γ for all $\gamma\in\Gamma$. Note that if $\Gamma\subseteq\Delta$ then what is left invariant under Δ is also left invariant under Γ.

We have hitherto applied the notion of invariance to relations. It is easy to generalize this notion so as to be applicable to structures. Let γ be a correspondence from A to A. A relational structure $\langle A,\langle\pi_i\rangle_{i<\alpha}\rangle$ is invariant under γ if for all $i<\alpha$, π_i is invariant under γ. (For relational structures, see 2.8.)

2.8. Relational structures

Let α be an ordinal. A function with the set $\{i \mid i < \alpha\}$ as domain is called an α-*termed sequence*, or simply, an α-*sequence*. If f is an α-sequence we often write f_i instead of $f(i)$, and f_i is called the ith *term* of the sequence f. Whenever f is an α-sequence we write f as

$$\langle f_i : i < \alpha \rangle$$

or as

$$\langle f_i \rangle_{i < \alpha} .$$

The *concatenation* of the two α-sequences $f = \langle f_i \rangle_{i < \alpha}$ and the β-sequence $g = \langle g_i \rangle_{i < \beta}$ is the $(\alpha+\beta)$-sequence $h = \langle h_i \rangle_{i < (\alpha+\beta)}$ such that for every $i < \alpha+\beta$

$$h_i = \begin{cases} f_i & \text{if } i < \alpha \\ g_\gamma & \text{if } i = \alpha+\gamma \end{cases}$$

We denote the concatenation of f and g by $f^\wedge g$. Note that concatenation of ordinal-termed sequences is associative, i.e. $(f^\wedge g)^\wedge h = f^\wedge (g^\wedge h)$.

A *relational structure* **A**, or simply a *structure* **A**, is an ordered pair $\langle A, \Pi \rangle$ where A is a nonvoid set and Π an ordinal-termed sequence whose terms are relations on A. A is called the *base set* of **A**. If $\Pi = \langle \pi_i \rangle_{i < \alpha}$ then we often write **A** as

$$\langle A, \pi_i \rangle_{i < \alpha} .$$

If $\Pi = \langle \pi_0, ..., \pi_n \rangle$ we write $\langle A, \pi_0, ..., \pi_{n-1} \rangle$ for **A**.

Let $\mathbf{A} = \langle A, \pi_i \rangle_{i < \alpha}$ and $\mathbf{B} = \langle B, \rho_i \rangle_{i < \beta}$ be structures such that A=B. The concatenation of **A** and **B** is the structure $\mathbf{C} = \langle C, \sigma_i \rangle_{i < (\alpha+\beta)}$ where C=A and $\langle \sigma_i \rangle_{i < (\alpha+\beta)}$ the concatenation of $\langle \pi_i \rangle_{i < \alpha}$ and $\langle \rho_i \rangle_{i < \beta}$. Hence, for every $i < \alpha+\beta$

$$\sigma_i = \begin{cases} \pi_i & \text{if } i < \alpha \\ \rho_\gamma & \text{if } i = \alpha+\gamma \end{cases}$$

We denote the concatenation of the structure **A** and **B** by $\mathbf{A}^\wedge \mathbf{B}$ or simply **AB**, i.e. by the juxtaposition of "A" and "B". Since concatenation of ordinal-termed sequences is associative this is also the case for structures; i.e.

$$(AB)C = A(BC),$$

and we omit the parentheses and write just **ABC**.

In section 2.6 the meaning of $\gamma[\rho]$ was defined. We generalize this definition to structures. If $A=\langle A,\langle\pi_i\rangle_{i<\alpha}\rangle$ and γ a correspondence from A to A, then

$$\gamma[A] = \langle\; \gamma[A]\;,\langle\gamma[\pi_i]\rangle_{i<\alpha}\;\rangle.$$

The following lemma is almost obvious.

Lemma 2.8.1: If **A** and **B** are relational structures with the same base set A and φ is a bijection from A to A', then

$$\varphi[AB] = \varphi[A]^\wedge\varphi[B].$$

If $\langle A,\pi_i\rangle_{i<\alpha}$ and $\langle B,\rho_i\rangle_{i<\alpha}$ are relational structures such that for all i, $i<\alpha$, π_i and ρ_i have the same arity, then we say that $\langle A,\pi_i\rangle_{i<\alpha}$ and $\langle B,\rho_i\rangle_{i<\alpha}$ are *similar* relational structures. By the [similarity] *type* of $\langle A,\pi_i\rangle_{i<\alpha}$ we understand the sequence $\langle n_i\rangle_{i<\alpha}$ such that for all i, $i<\alpha$, π_i is an n_i-ary relation.

Suppose $A=\langle A,\pi_i\rangle_{i<\alpha}$. Then the restriction of **A** to the set B, which is denoted A/B, is the relational structure $\langle A\cap B,\pi_i/B\rangle_{i<\alpha}$

2.9 Isomorphisms and homomorphisms

In this section the important notions of an isomorphism and a homomorphism are introduced.

Let ρ be an n-ary relation in A and σ an n-ary relation in B. A *homomorphism* of (or on) $\langle A,\rho\rangle$ *into* $\langle B,\sigma\rangle$ is a function φ on A into B such that for all $a_1,...,a_n\in A$,

$$\rho(a_1,...,a_n) \text{ iff } \sigma(\varphi(a_1),...,\varphi(a_n)).$$

A *homomorphism* of (or on) $\langle A,\rho\rangle$ *onto* (or *to*) $\langle B,\sigma\rangle$ is a homomorphism φ on $\langle A,\rho\rangle$ into $\langle B,\sigma\rangle$ such that φ is onto B. An *isomorphism* of (or on) $\langle A,\rho\rangle$ *into* $\langle B,\sigma\rangle$ is a homomorphism φ on $\langle A,\rho\rangle$ into $\langle B,\sigma\rangle$ such that φ is a one-to-one function. An *isomorphism* of (or on) $\langle A,\rho\rangle$ *onto* (or *to*) $\langle B,\sigma\rangle$ is an isomorphism φ on $\langle A,\rho\rangle$ into $\langle B,\sigma\rangle$ such that φ is onto B. If there is an isomorphism φ on $\langle A,\rho\rangle$ onto $\langle B,\sigma\rangle$ we say that $\langle A,\rho\rangle$ and $\langle B,\sigma\rangle$ are isomorphic and we write $\langle A,\rho\rangle\cong\langle B,\sigma\rangle$. An isomorphism φ on $\langle A,\rho\rangle$ onto $\langle A,\rho\rangle$ is called an *automorphism* of $\langle A,\rho\rangle$. An automorphism φ of $\langle A,\rho\rangle$ is thus a one-to-one function on A onto A such that for all $a_1,...,a_n\in A$,

$$\rho(a_1,...,a_n) \text{ iff } \rho(\varphi(a_1),...,\varphi(a_n)).$$

The concept of a homomorphism introduced above is the one used in measurement theory of relations. (See Scott and Suppes, 1958, p. 114; Kanger, 1963; Kanger, 1972, p. 4; Krantz et. al., 1971, p. 9; Pfanzagl, 1971, p. 23 and Roberts, 1979, p. 52.) It is also *one* of the notions of a homomorphism of a relational structure used in mathematics. But there seems to be at least one other notion of a homomorphism in mathematics. In universal algebra, a homomorphism of $\langle A,\rho \rangle$ into $\langle B,\sigma \rangle$ is often understood to be a function φ on A into B such that for all $a_1,...,a_n \in A$,

$$\rho(a_1,...,a_n) \Rightarrow \sigma(\varphi(a_1),...,\varphi(a_n)).$$

(See for example Grätzer, 1979, p. 224.)

We have so far only defined isomorphisms and homomorphisms for relational structures containing *one* relation. It is straightforward to generalize to structures containing an arbitrary number of relations. Let $A=\langle A,\langle \rho_i \rangle_{i<\alpha} \rangle$ and $B=\langle B,\langle \sigma_i \rangle_{i<\alpha} \rangle$ be similar relational systems, i.e. ρ_i and σ_i have the same arity for all $i<\alpha$. An isomorphism of (on) A onto B is for example a function φ on A onto B such that φ is an isomorphism on $\langle A,\rho_i \rangle$ onto $\langle B,\sigma_i \rangle$ for all $i<\alpha$. And analogously for the other notions of morphisms. An *isomorphism* of (or on) $\langle A,\rho_i \rangle_{i<\alpha}$ to $\langle B,\sigma_i \rangle_{i<\alpha}$ is thus a one-to-one function φ on A onto B such that for all $i<\alpha$ and all $a_1,...,a_{n_i} \in A$,

$$\rho_i(a_1,...,a_{n_i}) \text{ iff } \sigma_i(\varphi(a_1),...,\varphi(a_{n_i})).$$

We introduce the following notational convention The set of homomorphism of A *into* B is denoted by $\mathbb{H}(A,B)$. The set of isomorphisms of A *onto* B is denoted by $\mathbb{I}(A,B)$. The set of automorphisms of A is denoted by $\mathbb{I}(A)$ rather than by $\mathbb{I}(A,A)$. Note that $\mathbb{I}(A,B) \subseteq \mathbb{H}(A,B)$.

Note that $\varphi \in \mathbb{I}(\langle A,\langle \rho_i \rangle_{i<\alpha} \rangle, \langle B,\langle \sigma_i \rangle_{i<\alpha} \rangle)$ iff $\varphi \in \mathbb{I}(\langle A,\rho_i \rangle, \langle B,\sigma_i \rangle)$ for all $i<\alpha$ iff $\varphi \in \cap_{i<\alpha} \mathbb{I}(\langle A,\rho_i \rangle, \langle B,\sigma_i \rangle)$.

In the following lemma a number of simple results on isomorphisms are collected together, some of which will be frequently used in later chapters. First a reminder. If $A=\langle A,\langle \pi_i \rangle_{i<\alpha} \rangle$ and γ is a correspondence from A to A, then $\gamma[A] = \langle \gamma[A] ,\langle \gamma[\pi_i] \rangle_{i<\alpha} \rangle$.

Theorem 2.9.1: Suppose A,B,C and D are similar relational structures.

(1) $\varphi \in \mathbb{I}(A,B)$ iff $\varphi^{-1} \in \mathbb{I}(B,A)$

(2) $\varphi \in \mathbb{I}(A,B)$ & $\psi \in \mathbb{I}(B,C) \Rightarrow \psi \circ \varphi \in \mathbb{I}(A,C)$

(3) $\varphi \in \mathbb{I}(A,B)$ iff $\varphi \in \mathbb{B}i(A,B)$ & $\varphi[A]=B$.

(4) If $\varphi \in \mathbb{B}i(A,B)$, then $\varphi[A]=B$ iff $\varphi^{-1}[B]=A$.

(5) If $\varphi \in \mathbb{B}i(A,C)$ and $\psi \in \mathbb{B}i(C,B)$, then $\psi[\varphi[A]]=B$ iff $(\psi \circ \varphi)[A]=B$.

(6) If $\varphi \in \mathbb{B}i(A,C)$ and $\psi \in \mathbb{B}i(B,C)$, then $\varphi[A]=\psi[B]$ iff $(\psi^{-1} \circ \varphi)[A]=B$.

(7) $\varphi \in \mathbb{I}(\mathbf{A},\mathbf{B})$ & $\psi \in \mathbb{I}(\mathbf{A},\mathbf{C})$ $\Rightarrow$ $(\psi o \varphi^{-1}) \in \mathbb{I}(\mathbf{B},\mathbf{C})$.

(8) $\varphi \in \mathbb{I}(\mathbf{A},\mathbf{B})$ & $\psi \in \mathbb{I}(\mathbf{C},\mathbf{D})$ & $(\psi o \varphi^{-1}) \in \mathbb{I}(\mathbf{B},\mathbf{D})$ $\Rightarrow$ $\mathbf{A}=\mathbf{C}$.

(9)   If $\varphi \in \mathbb{B}i(A,C)$ and $\psi \in \mathbb{B}i(B,C)$, then $\varphi[A] \neq \psi[B]$ iff $(\psi^{-1} o \varphi) \notin \mathbb{I}(\mathbf{A},\mathbf{B})$.

(10) If $\varphi \in \mathbb{I}(\mathbf{A},\mathbf{B})$ and $\psi \notin \mathbb{I}(\mathbf{A},\mathbf{B})$, then $(\psi^{-1} o \varphi) \notin \mathbb{I}(\mathbf{A})$.

Proof: We prove only (3), (8) and (9).

(3) We prove first the proposition for the case that $\mathbf{A}=\langle A,\rho \rangle$ and $\mathbf{B}=\langle B,\sigma \rangle$.
(I) Suppose $\varphi \in \mathbb{I}(\mathbf{A},\mathbf{B})$. Then $\varphi \in \mathbb{B}i(A,B)$ and thus $\varphi[A]=B$. It remains to prove that $\varphi[\rho]=\sigma$. Suppose that $\langle b_1,...,b_n \rangle \in \varphi[\rho]$. Then there are $a_1,...,a_n \in A$ such that $\rho(a_1,...,a_n)$ and $\varphi(a_1)=b_1,...,\varphi(a_n)=b_n$. Since $\varphi \in \mathbb{I}(\mathbf{A},\mathbf{B})$ it follows that $\sigma(\varphi(a_1),...,\varphi(a_n))$ and thus $\sigma(b_1,...,b_n)$. We have thus proved that $\varphi[\rho] \subseteq \sigma$. Suppose now that $\sigma(b_1,...,b_n)$. Then there are $a_1,...,a_n \in A$ such that $\varphi(a_1)=b_1,...,\varphi(a_n)=b_n$. Since $\varphi \in \mathbb{I}(\mathbf{A},\mathbf{B})$, $\sigma(b_1,...,b_n)$ implies that $\rho(a_1,...,a_n)$, and this implies that $\langle \varphi(a_1),...,\varphi(a_n) \rangle \in \varphi[\rho]$. This proves that $\sigma \subseteq \varphi[\rho]$.
(II) Suppose that $\varphi \in \mathbb{B}i(A,B)$ and $\varphi[A]=\mathbf{B}$. Suppose further that $\rho(a_1,...,a_n)$. Then $\langle \varphi(a_1),...,\varphi(a_n) \rangle \in \varphi[\rho]$. Since $\varphi[\rho]=\sigma$, it follows that $\langle \varphi(a_1),...,\varphi(a_n) \rangle \in \sigma$. We have thus proved that $\rho(a_1,...,a_n)$ implies $\sigma(\varphi(a_1),...,\varphi(a_n))$. Suppose now that $\sigma(\varphi(a_1),...,\varphi(a_n))$. Since $\varphi[\rho]=\sigma$, $\langle \varphi(a_1),...,\varphi(a_n) \rangle \in \varphi[\rho]$. Then there are $x_1,...,x_n \in A$ such that $\rho(x_1,...,x_n)$ and $\varphi(a_1)=x_1,...,\varphi(a_n)=x_n$ Since φ is a one-to-one function it follows that $a_1=x_1,...,a_n=x_n$. Hence, $\rho(a_1,...,a_n)$. We have thus proved that $\sigma(\varphi(a_1),...,\varphi(a_n))$ implies $\rho(a_1,...,a_n)$. Therefore, $\varphi \in \mathbb{I}(\mathbf{A},\mathbf{B})$.

We now prove the general case. Suppose $\mathbf{A}=\langle A,\langle \rho_i \rangle_{i<\alpha} \rangle$ and $\mathbf{B}=\langle B,\langle \sigma_i \rangle_{i<\alpha} \rangle$. $\varphi \in \mathbb{I}(\mathbf{A},\mathbf{B})$ iff $\varphi \in \mathbb{I}(\langle A,\rho_i \rangle,\langle B,\sigma_i \rangle)$ for all $i<\alpha$ iff $\varphi \in \mathbb{B}i(A,B)$ & $\varphi[\langle A,\rho_i \rangle]=\langle B,\sigma_i \rangle$ for all $i<\alpha$ iff $\varphi \in \mathbb{B}i(A,B)$ & $\langle \varphi[A],\varphi[\rho_i] \rangle=\langle B,\sigma_i \rangle$ for all $i<\alpha$ iff $\langle \varphi[A],\langle \varphi[\rho_i] \rangle_{i<\alpha} \rangle=\langle B,\langle \sigma_i \rangle_{i<\alpha} \rangle$ iff $\varphi \in \mathbb{B}i(A,B)$ & $\varphi[A]=\mathbf{B}$.

(8) $(\psi o \varphi^{-1}) o \varphi \in \mathbb{I}(\mathbf{A},\mathbf{D})$ according to (2), which implies that $\psi \in \mathbb{I}(\mathbf{A},\mathbf{D})$. Since $\psi^{-1} o \psi \in \mathbb{I}(\mathbf{A},\mathbf{C})$, $\iota_A \in \mathbb{I}(\mathbf{A},\mathbf{C})$ and thus according to (3), $\iota_A[A]=C$. This implies $\mathbf{A}=\mathbf{C}$.

(9) Suppose $\varphi \in \mathbb{B}i(A,C)$ and $\psi \in \mathbb{B}i(B,C)$. Then $(\psi^{-1} o \varphi) \in \mathbb{B}i(A,B)$. Using (3) and (6) we get the following equivalences. $(\psi^{-1} o \varphi) \notin \mathbb{I}(\mathbf{A},\mathbf{B})$ iff $(\psi^{-1} o \varphi)[A] \neq B$ iff $\varphi[A] \neq \psi[B]$. ♦

Note the connection between the notion of invariance and the notion of an automorphism. Suppose $\mathbf{A}=\langle A,\Pi \rangle$, where Π is a sequence of relations in A, and $\varphi \in \mathbb{B}i(A)$. $\mathbf{A}=\langle A,\Pi \rangle$ is invariant under φ iff for all $\rho \in \Pi$, where ρ is n-ary, and all $a_1,...a_n \in A$,

$$\rho(a_1,...a_n) \text{ iff } \rho(\varphi(a_1),...,\varphi(a_n)),$$

i.e., iff $\varphi \in \mathbb{I}(A)$. If $\mathbf{A}$ and $\mathbf{B}$ are structures with the same base set A, then $\mathbf{B}$ is automorphism-invariant relative to $\mathbf{A}$ iff $\mathbf{B}$ is invariant under the set of automorphisms of $\mathbf{A}$, i.e. iff $\mathbb{I}(\mathbf{B}) \supseteq \mathbb{I}(\mathbf{A})$.

Theorem 2.9.2: Suppose that $\mathbf{A}$ and $\mathbf{B}$ are similar relational structures and that $\mathbf{C}$ and $\mathbf{D}$ are similar relational structures and that A=C and B=D. Then
$$\mathbb{I}(\mathbf{A},\mathbf{B}) \cap \mathbb{I}(\mathbf{C},\mathbf{D}) = \mathbb{I}(\mathbf{AC},\mathbf{BD}).$$
Proof: Suppose $\mathbf{A}=\langle A,\langle\rho_i\rangle_{i<\alpha}\rangle$, $\mathbf{B}=\langle B,\langle\sigma_i\rangle_{i<\alpha}\rangle$, $\mathbf{C}=\langle A,\langle\pi_i\rangle_{i<\beta}\rangle$ and $\mathbf{D}=\langle B,\langle\tau_i\rangle_{i<\beta}\rangle$. Suppose further that ρ_i and σ_i have the same arity for all $i<\alpha$, and that π_i and τ_i have the same arity for all $i<\beta$. Then $\mathbf{AC}=\langle A,\langle\zeta_i\rangle_{i<\alpha+\beta}\rangle$ where

$$\zeta_i = \begin{cases} \rho_i & \text{if} \quad i<\alpha \\ \pi_j & \text{if} \quad i=\alpha+j \end{cases}$$

and $\mathbf{BD}=\langle B,\langle\eta_i\rangle_{i<\alpha+\beta}\rangle$ where

$$\eta_i = \begin{cases} \sigma_i & \text{if} \quad i<\alpha \\ \tau_j & \text{if} \quad i=\alpha+j \end{cases}$$

and ζ_i and η_i have the same arity for all $i<\alpha+\beta$. This implies that $\mathbf{AC}$ and $\mathbf{BD}$ are similar.
$$\varphi \in (\mathbb{I}(\mathbf{A},\mathbf{B}) \cap \mathbb{I}(\mathbf{C},\mathbf{D})) \text{ iff}$$
$$\text{iff} \quad \varphi \in \mathbb{I}(\langle A,\rho_i\rangle,\langle B,\sigma_i\rangle) \text{ for all } i<\alpha \text{ and } \varphi \in \mathbb{I}(\langle C,\pi_i\rangle,\langle D,\tau_i\rangle) \text{ for all } i<\beta \text{ iff}$$
$$\text{iff} \quad \varphi \in \mathbb{I}(\langle A,\zeta_i\rangle,\langle B,\eta_i\rangle) \text{ for all } i<\alpha+\beta \text{ iff}$$
$$\text{iff} \quad \mathbb{I}(\mathbf{AC},\mathbf{BD}). \quad \blacklozenge$$

Corollary 2.9.3: Suppose that $\mathbf{A}$ and $\mathbf{C}$ are relational structures with the same base set A. Then
$$\mathbb{I}(\mathbf{A}) \cap \mathbb{I}(\mathbf{C}) = \mathbb{I}(\mathbf{AC}).$$

2.10 Congruence relations

In this section the important notion of a congruence relation is introduced. The following definition is essentially the one given in Pfanzagl (1971) p. 20.

Let ρ be an n-ary relation on A. An equivalence relation $\approx$ on A is a *congruence relation* for ρ if $\approx$ has the following substitution property: If for all i, $1\leq i\leq n$, $a_i\approx b_i$, then $\rho(a_1,...,a_n)$ iff $\rho(b_1,...,b_n)$.

It is easy to see that, since $\approx$ is an equivalence relation and thus symmetric, the substitution property can equivalently be formulated as follows: If for all i,$1\leq i\leq n$, $a_i\approx b_i$, then $\rho(a_1,...,a_n)$ implies $\rho(b_1,...,b_n)$. Note also that if $\approx$ satisfies the substitution property and $\tau\subseteq\approx$, then τ satisfies the substitution property too.

This definition of a congruence relation for a relation can easily be generalized to a definition of a congruence relation for a relational structure. First a reminder. As was said in section 2.8, if $\Pi=\langle\pi_i\rangle_{i<\alpha}$ we use $\rho\in\Pi$ to denote that $\rho=\pi_j$ for some $j<\alpha$.

Let $\langle A,\Pi\rangle$ be a relational structure. An equivalence relation $\approx$ on A is a *congruence relation* for $\langle A,\Pi\rangle$ if for all $\rho\in\Pi$, $\approx$ on A is a congruence relation for ρ.

The notion of a congruence relation is closely related to the notion of invariance. An equivalence relation on A is a correspondence on A onto A. In section 2.7 we defined what is meant by invariance under a correspondence: An n-ary relation ρ on A is *invariant under* the correspondence γ from A to A if for all $a_1,...,a_n,b_1,...,b_n\in A$ such that $a_1\gamma b_1,,...,a_n\gamma b_n$:

$$\rho(a_1,...,a_n) \Leftrightarrow \rho(b_1,...,b_n).$$

Hence, γ is a congruence relation for ρ iff γ is an equivalence relation and the relation ρ is invariant under γ.

Theorem 2.10.1: The equivalence relation $\approx$ on A is a congruence relation for the relation ρ on A iff $\approx[\rho]=\rho$.

Proof: Suppose $\approx$ is an equivalence relation on A. The symmetry of $\approx$ implies that $\approx=\approx^{-1}$. The reflexivity of $\approx$ implies that $\approx[A]=\approx^{-1}[A]=A$. $\approx$ is a congruence relation for ρ iff ρ is invariant under $\approx$, which according to 2.7.7 is equivalent to $\approx[\rho]=\rho/\approx[A]$ and $\approx^{-1}[\rho]=\rho/\approx^{-1}[A]$, which holds, since $\approx$ is an equivalence relation, iff $\approx[\rho]=\rho$. $\blacklozenge$

2.10.1 can be generalized to structures and we do this in a corollary to 2.10.1. First a reminder. If $A=\langle A,\langle\pi_i\rangle_{i<\alpha}\rangle$ and γ is a correspondence from A to A, then $\gamma[A]=\langle\gamma[A],\langle\gamma[\pi_i]\rangle_{i<\alpha}\rangle$.

Corollary 2.10.2: The equivalence relation $\approx$ on A is a congruence relation for the structure A with base set A iff $\approx[A]=A$.

The similarity and difference between an automorphism and a congruence relation is worth noting. From 2.7.13 follows that γ is an automorphism of A iff γ is a bijection on A and $\gamma[\rho]=\rho$. And 2.10.1 shows that γ is a congruence relation for A iff γ is an equivalence relation on A and $\gamma[\rho]=\rho$.

It is important to note what this definition of congruence relations implies for operations. An operation is a special kind of relation. Suppose ϖ is an n-ary operation on A. Define the n+1-ary relation σ by

$$\sigma(a_1,...,a_n,a_{n+1}) \text{ iff } \varpi(a_1,...,a_n)=a_{n+1}.$$

Since ϖ is an operation on A it follows that $\sigma(a_1,...,a_n,x)$ and $\sigma(a_1,...,a_n,y)$ implies x=y , and for all $a_1,...,a_n \in A$ there is z$\in$ A such that $\sigma(a_1,...,a_n,z)$. The n-ary operation ϖ can thus be regarded as an n+1-ary relation σ on A where σ is a relation of a special kind.

Theorem 2.10.3: Suppose ϖ is an n-ary operation on A and σ the n+1-ary relation on A defined by

$$\sigma(a_1,...,a_n,a_{n+1}) \text{ iff } \varpi(a_1,...,a_n)=a_{n+1}.$$

The equivalence relation $\approx$ on A is a congruence relation for σ iff the following two condition holds:

(1) $a_i \approx b_i$ for all i, $1 \leq i \leq n$, implies

$$\varpi(a_1,...,a_n)=\varpi(b_1,...,b_n).$$

(2) if x belongs to the range of ϖ and x$\approx$y then x=y.

Proof: (I) Suppose $\approx$ is a congruence relation for σ. We first prove (1). Suppose $a_i \approx b_i$ for all i, $1 \leq i \leq n$. Since $\approx$ is reflexive ,

$$\varpi(a_1,...,a_n) \approx \varpi(a_1,...,a_n).$$

From the fact that $\approx$ is a congruence relation for σ, it follows that

$$\sigma(a_1,...,a_n,\varpi(a_1,...,a_n)) \text{ iff } \sigma(b_1,...,b_n,\varpi(a_1,...,a_n)).$$

This implies, using the definition of σ, that

$$\varpi(a_1,...,a_n)=\varpi(b_1,...,b_n).$$

We so prove (2). Suppose there is $a_1,...,a_n \in A$ such that $\varpi(a_1,...,a_n)=x$ and x$\approx$y. Thus, $\sigma(a_1,...,a_n,x)$ and since $\approx$ is a congruence relation for σ and x$\approx$y, then $\sigma(a_1,...,a_n,y)$. From this and the definition of σ it follows that $\varpi(a_1,...,a_n)=y$, and, since ϖ is a function, x=y.

(II) Suppose $\approx$ is an equivalence relation on A and that (1) and (2) holds. Suppose further that $a_i \approx b_i$ for all i, $1 \leq i \leq n+1$, and $\sigma(a_1,...,a_n,a_{n+1})$. Thus, $\varpi(a_1,...,a_n)=a_{n+1}$, and according to (1), $\varpi(b_1,...,b_n)=a_{n+1}$. From this and the definition of σ it follows

$\sigma(b_1,...,b_n,a_{n+1})$. a_{n+1} belongs to the range of ϖ and $a_{n+1}\approx b_{n+1}$ and thus according to (2) $a_{n+1}=b_{n+1}$. This implies that $\sigma(b_1,...,b_n,b_{n+1})$. Hence, $\approx$ is a congruence relation for σ. ◆

2.10.3 shows that the definition of a congruence relation stated above, when applied to an operation, gives a condition different from the usual one defining congruence relations in algebra. In algebra, the equivalence relation $\approx$ on A is a congruence relation for the n-ary operation ϖ on A iff for all i,$1\leq i\leq n$, $a_i\approx b_i$, implies that

$$\varpi(a_1,...,a_n)\approx\varpi(b_1,...,b_n).$$

In this definition $\varpi(a_1,...,a_n)$ and $\varpi(b_1,...,b_n)$ stand in the relation $\approx$ to each other while in the definition giving above = holds between them. This is important to remember, since the notion of a congruence relation adopted here will have properties differing from those of congruence relations in algebra. The notion of a congruence relation is connected with the notion of a homomorphism, and as was pointed out in section 2.9 we also use a notion of a homomorphism which differs from the one often used in algebra.

Let ϖ be an n-ary operation on A and ϖ' an n-ary operation on A'. In algebra, a homomorphism of $\langle A,\varpi\rangle$ onto $\langle A',\varpi'\rangle$ is often understood to mean a function φ on A into A' such that for all $a_1,...,a_n\in A$,

$$\varpi'(\varphi(a_1),...,\varphi(a_n)) = \varphi(\varpi(a_1,..,a_n)).$$

The following theorem clarifies the relation between this definition of a homomorphism and the one adopted in section 2.9.

Theorem 2.10.4: Suppose ϖ is an n-ary operation on A and σ the $n+1$-ary relation on A defined by for all $a_1,...,a_n,a_{n+1}\in A$,

$$\sigma(a_1,...,a_n,a_{n+1}) \text{ iff } \varpi(a_1,...,a_n)=a_{n+1}.$$

Suppose ϖ' is an n-ary operation on A' and σ' the $n+1$-ary relation on A' defined by for all $a_1,...,a_n,a_{n+1}\in A'$,

$$\sigma'(a_1,...,a_n,a_{n+1}) \text{ iff } \varpi'(a_1,...,a_n)=a_{n+1}.$$

Suppose further that φ is a function on A into A'. Then the following two conditions are equivalent.

(1) For all $a_1,...,a_n,a_{n+1}\in A$:

$$\sigma(a_1,...,a_n,a_{n+1}) \Rightarrow \sigma'(\varphi(a_1),...,\varphi(a_n),\varphi(a_{n+1})).$$

(2) For all $a_1,...,a_n \in A$:
$$\varpi'(\varphi(a_1),...,\varphi(a_n)) = \varphi(\varpi(a_1,..,a_n)).$$
Proof: Suppose (1). From the definition of σ it follows that
$$\sigma(a_1,...,a_n,\varpi(a_1,...,a_n)),$$
which according to (1) implies $\sigma'(\varphi(a_1),...,\varphi(a_n),\varphi(\varpi(a_1,..,a_n)))$. Thus, from the definition of σ' it follows that
$$\varpi'(\varphi(a_1),...,\varphi(a_n)) = \varphi(\varpi(a_1,..,a_n)).$$
(II) Suppose (2). Suppose further that $\sigma(a_1,...,a_n,a_{n+1})$. Thus, $\varpi(a_1,..,a_n) = a_{n+1}$. According to (2),
$$\varpi'(\varphi(a_1),...,\varphi(a_n)) = \varphi(a_{n+1}),$$
i.e. $\sigma'(\varphi(a_1),...,\varphi(a_n),\varphi(a_{n+1}))$. ♦ ·

An equivalence relation $\approx_2$ is said to be *coarser* than an equivalence relation $\approx_1$ if $\approx_2 \supseteq \approx_1$, or equivalently, if $a\approx_1 b$ implies $a\approx_2 b$. (See Pfanzagl, 1971, p. 21.) According to this definition a congruence relation is coarser than itself. This might seem strange, and we can of course modify the definition so that it does not hold. However, for convenience we accept Pfanzagl's definition here. If $\approx_2$ is coarser than $\approx_1$, then $\approx_1$ is *finer* than $\approx_2$. The term "weaker" is sometimes used with the same meaning as the term "coarser", i.e. an equivalence relation $\approx_2$ is *weaker* than another equivalence relation $\approx_1$ iff $a\approx_1 b$ implies $a\approx_2 b$. If $\approx_2$ is weaker than $\approx_1$, then $\approx_1$ is *stronger* than $\approx_2$.

Let $A=\langle A, \langle \pi_i \rangle_{i<\alpha} \rangle$ be a relational structure where π_i is n_i-ary. In Scott and Suppes (1958) p. 118 a unique equivalence relation $\sim$ for A is defined with the following meaning: a and b stand in the relation just when they are perfect substitutes for each other with respect to all the relations π_i. The formal definition of $\sim$ is somewhat complicated and runs as follows.

$a\sim b$ iff for each $i<\alpha$ and each pair $\langle z_1,...,z_{n_i} \rangle, \langle w_1,...,w_{n_i} \rangle$ of n_i-termed sequences of elements of A, if $z_j \neq w_j$ implies $\{z_j,w_j\}=\{a,b\}$ for $j=1,...n_i$, then
$$\pi_i(z_1,...,z_{n_i}) \text{ iff } \pi_i(w_1,...,w_{n_i}).$$

Theorem 2.10.5: $\sim$ is the coarsest congruence relation for $A=\langle A,\Pi \rangle$, i.e. $\sim$ is a congruence relation for A and if $\approx$ is a congruence relation for A then $\approx \subseteq \sim$.
Proof: (I) We first prove that $\sim$ is a congruence relation for A. Suppose $\rho \in \Pi$ and n-ary. Suppose further that $a_i \sim b_i$ for all i, $1 \leq i \leq n$, and that $\rho(a_1,...,a_n)$. Let $z_j=a_j$ for all $j=1,...,n$. Further, let $w_1=b_1$ and $w_j=a_j$ for all $j=2,...,n$. $\langle z_1,...,z_n \rangle$ and

$\langle w_1,...,w_n \rangle$ are n-termed sequences of elements of A such that $z_j \neq w_j$ implies $\{z_j,w_j\}=\{a_1,b_1\}$ for $j=1,...n$. From the definition of ~ it then follows that

$$\rho(a_1,a_2,...,a_n) \text{ iff } \rho(b_1,a_2,...,a_n).$$

In an analogous way we prove that

$$\rho(b_1,a_2,...,a_n) \text{ iff } \rho(b_1,b_2,...,a_n).$$

We proceed in this way until we have proved that

$$\rho(a_1,...,a_n) \text{ iff } \rho(b_1,...,b_n).$$

(II) We now prove that ~ is the coarsest congruence relation for **A**. Suppose $\approx$ is a congruence relation for **A** and that $a \approx b$. Let $\langle z_1,...,z_n \rangle$ and $\langle w_1,...,w_n \rangle$ be n-termed sequences of elements of A such that $z_j \neq w_j$ implies $\{z_j,w_j\}=\{a,b\}$ for $j=1,...n$. Then either $z_j=w_j$ or $\{z_j,w_j\}=\{a,b\}$. In both cases $z_j \approx w_j$ and since $\approx$ is a congruence relation for **A**,

$$\rho(z_1,...,z_n) \text{ iff } \rho(w_1,...,w_n)$$

where $\rho \in \Pi$ and is n-ary. From this it follows that $a \sim b$. We have thus proved that if $\approx$ is a congruence relation for **A**, then $\approx \subseteq \sim$. ~ is therefore coarser than every congruence relation for **A**. ♦

The preceding theorem resembles to a large extent Theorem 1.4.5 in Pfanzagl (1971) p. 21.

It is often useful to denote the coarsest congruence relation for **A** by $\sim_A$ so that the reference to the structure is made explicit.

The set of all congruence relations for **A** is denoted by $\mathbb{K}(A)$. We can define a notion of congruence invariance which is analogous to the notion of automorphism invariance defined in section 2.9. Suppose **A** and **B** are relational structures with the same base set, i.e. A=B. Let us say that **B** is *congruence-invariant* relative to **A** if **B** is invariant under $\mathbb{K}(A)$. That **B** is invariant under the congruence relation $\approx$ for **A** is, as we have seen above, equivalent to $\approx$ being a congruence relation for **B**. **B** is thus congruence-invariant relative to **A** iff $\mathbb{K}(B) \supseteq \mathbb{K}(A)$. For comparison, remember that **B** is automorphism-invariant relative to **A** if $\mathbb{I}(B) \supseteq \mathbb{I}(A)$.

Theorem 2.10.6: $\mathbb{K}(B) \supseteq \mathbb{K}(A)$ iff $\sim_B \supseteq \sim_A$.

Proof: If $\mathbb{K}(B) \supseteq \mathbb{K}(A)$ then $\sim_A \in \mathbb{K}(B)$ and thus $\sim_A \subseteq \sim_B$. Suppose now that $\sim_B \supseteq \sim_A$ and that $\approx \in \mathbb{K}(A)$. Then $\approx \subseteq \sim_A \subseteq \sim_B$ so $\approx$ satisfies the substitution property for **B**, and since $\approx$ is an equivalence relation, $\approx \in \mathbb{K}(B)$. ♦

2.10.6 shows that **B** is congruence-invariant relative to **A** iff the coarsest congruence relation for **B** is coarser than the coarsest congruence relation for **A**.

The next three propositions deal with the connection between congruence relations for **AB** and for **A** and **B**.

Lemma 2.10.7: Suppose **A** and **B** are relational structures with the same base set A. $\approx$ is a congruence relation for **AB** iff $\approx$ is a congruence relation for **A** and for **B**, i.e. $\mathbb{K}(\mathbf{AB}) = \mathbb{K}(\mathbf{A}) \cap \mathbb{K}(\mathbf{B})$.

Proof: Suppose $\mathbf{A} = \langle A, \Pi \rangle$ and $\mathbf{B} = \langle B, \Sigma \rangle$. Then $\mathbf{AB} = \langle A, \Pi \wedge \Sigma \rangle$. If $\approx$ is a congruence relation for **AB** then $\approx$ is a congruence relation for every $\rho \in \Pi$ and every $\sigma \in \Sigma$ and thus a congruence relation for **A** and for **B**. And if $\approx$ is a congruence relation for **A** and for **B** then $\approx$ is also a congruence relation for every $\rho \in \Pi$ and every $\sigma \in \Sigma$, i.e. $\approx$ is a congruence relation for **AB**. ♦

Lemma 2.10.8: Suppose **A** and **B** are relational structures with the same base set A. If $\approx_1 \in \mathbb{K}(\mathbf{A})$ and $\approx_2 \in \mathbb{K}(\mathbf{B})$ then
$$\approx_1 \cap \approx_2 \in \mathbb{K}(\mathbf{AB}).$$
Proof: It is easy to prove that the intersection of two equivalence relations on the same set is an equivalence relation. Hence, $\approx_1 \cap \approx_2$ is an equivalence relation on A. Since $\approx_1$ has the substitution property for **A** the same holds for $\approx_1 \cap \approx_2$. Thus, $\approx_1 \cap \approx_2 \in \mathbb{K}(\mathbf{A})$. Analogously, we prove that $\approx_1 \cap \approx_2 \in \mathbb{K}(\mathbf{B})$. From 2.10.7 then it follows that $\approx_1 \cap \approx_2 \in \mathbb{K}(\mathbf{AB})$. ♦

Theorem 2.10.9: $\sim_{\mathbf{AB}} = \sim_{\mathbf{A}} \cap \sim_{\mathbf{B}}$.

Proof: From 2.10.5 it follows that $\sim_{\mathbf{A}} \cap \sim_{\mathbf{B}} \in \mathbb{K}(\mathbf{AB})$. Thus, since $\sim_{\mathbf{AB}}$ is the coarsest congruence relation for **AB**, $\sim_{\mathbf{A}} \cap \sim_{\mathbf{B}} \subseteq \sim_{\mathbf{AB}}$. According to 2.10.7, $\sim_{\mathbf{AB}} \in \mathbb{K}(\mathbf{A})$ and $\sim_{\mathbf{AB}} \in \mathbb{K}(\mathbf{B})$. Hence, $\sim_{\mathbf{AB}} \subseteq \sim_{\mathbf{A}}$ and $\sim_{\mathbf{AB}} \subseteq \sim_{\mathbf{B}}$. ♦

The following theorem shows the relation between the automorphisms of **A** and of $\sim_{\mathbf{A}}$.

Theorem 2.10.10: Suppose **A** and **B** are similar relational structures. If $\varphi \in \mathbb{I}(\mathbf{A}, \mathbf{B})$ then $\varphi \in \mathbb{I}(\sim_{\mathbf{A}}, \sim_{\mathbf{B}})$.

Proof: Suppose $\varphi \in \mathbb{I}(\mathbf{A}, \mathbf{B})$. Let $\mathbf{A} = \langle A, \rho_i \rangle_{i < \alpha}$ and $\mathbf{B} = \langle B, \sigma_i \rangle_{i < \alpha}$. Choose i arbitrarily such that $i < \alpha$. Let $\rho = \rho_i$ and $\sigma = \sigma_i$. Suppose further that $a \sim_{\mathbf{A}} b$ and that

$\langle z_1,...,z_n \rangle$ and $\langle w_1,...,w_n \rangle$ are arbitrary n-termed sequences of elements of B such that $z_j \neq w_j$ implies $\{z_j,w_j\}=\{\varphi(a),\varphi(b)\}$ for j=1,...,n. Consider $\langle \varphi^{-1}(z_1),...,\varphi^{-1}(z_n) \rangle$ and $\langle \varphi^{-1}(w_1),...,\varphi^{-1}(w_n) \rangle$. They are sequences of elements of A. Since φ is a bijection,

$$\varphi^{-1}(z_j) \neq \varphi^{-1}(w_j) \implies z_j \neq w_j \implies \{z_j,w_j\}=\{\varphi(a),\varphi(b)\} \implies \{\varphi^{-1}(z_j),\varphi^{-1}(w_j)\}=\{a,b\}.$$

From $a \sim_A b$ it then follows that

$$\rho(\varphi^{-1}(z_1),...,\varphi^{-1}(z_n)) \text{ iff } \rho(\varphi^{-1}(w_1),...,\varphi^{-1}(w_n)).$$

Since $\varphi \in \mathbb{I}(\mathbf{A},\mathbf{B})$,

$$\rho(\varphi^{-1}(z_1),...,\varphi^{-1}(z_n)) \text{ iff } \sigma(z_1,...,z_n)$$

and

$$\rho(\varphi^{-1}(w_1),...,\varphi^{-1}(w_n)) \text{ iff } \sigma(w_1,...,w_n).$$

Thus,

$$\sigma(z_1,...,z_n) \text{ iff } \sigma(w_1,...,w_n)$$

which implies $\varphi(a) \sim_B \varphi(b)$. We have thus proved that

$$a \sim_A b \implies \varphi(a) \sim_B \varphi(b).$$

It remains now to prove that

$$\varphi(a) \sim_A \varphi(b) \implies a \sim_A b.$$

Suppose that $\varphi(a) \sim_B \varphi(b)$ and that $\langle z_1,...,z_n \rangle$ and $\langle w_1,...,w_n \rangle$ are arbitrary n-termed sequences of elements of A such that $z_j \neq w_j$ implies $\{z_j,w_j\}=\{a,b\}$ for j=1,...,n. Consider $\langle \varphi(z_1),...,\varphi(z_n) \rangle$ and $\langle \varphi(w_1),...,\varphi(w_n) \rangle$. They are sequences of elements of B. Since φ is a bijection,

$$\varphi(z_j) \neq \varphi(w_j) \implies z_j \neq w_j \implies \{z_j,w_j\}=\{a,b\} \implies \{\varphi(z_j),\varphi(w_j)\}=\{\varphi(a),\varphi(b)\}.$$

From $\varphi(a) \sim_B \varphi(b)$ it then follows that

$$\sigma(\varphi(z_1),...,\varphi(z_n)) \text{ iff } \sigma(\varphi(w_1),...,\varphi(w_n)).$$

Since $\varphi \in \mathbb{I}(\mathbf{A},\mathbf{B})$,

$$\rho(z_1,...,z_n) \text{ iff } \sigma(\varphi(z_1),...,\varphi(z_n))$$

and

$$\rho(w_1,...,w_n) \text{ iff } \sigma(\varphi(w_1),...,\varphi(w_n)).$$

Thus,

$$\rho(z_1,...,z_n) \text{ iff } \rho(w_1,...,w_n)$$

which implies $a \sim_A b$. We have thus proved that

$$\varphi(a) \sim_B \varphi(b) \implies a \sim_A b.$$

This shows that $\varphi \in \mathbb{I}(\sim_A, \sim_B)$. ◆

Theorem 2.10.11: Suppose $A=\langle A,\rho \rangle$ and ρ an equivalence relation. Then $\sim_A = \rho$.

Proof: We first prove that ρ is a congruence for A. Suppose $a_1 \rho b_1$ and $a_2 \rho b_2$.

Suppose further that $a_1\rho a_2$. Since ρ is a symmetric relation we get $b_1\rho a_1$, $a_1\rho a_2$ and $a_2\rho b_2$, which implies $b_1\rho b_2$ because of the transitivity of ρ. Thus, $\rho\in\mathbb{K}(A)$. Suppose now that $\approx\in\mathbb{K}(A)$ and that $a\approx b$. Since $\approx$ and ρ are reflexive we get $a\approx a$ and $a\rho a$. $a\approx a$, $a\approx b$ and $a\rho a$ imply, since $\approx$ is a congruence relation for ρ, $a\rho b$. This shows that $\approx\subseteq\rho$. ρ is thus the coarsest congruence relation for ρ, i.e. $\sim_A=\rho$. ♦

Note that if $A=\langle A,=\rangle$ where $=$ is the identity relation, then $\mathbb{K}(A)=\{=\}$, since according to 2.10.11, $=$ is the coarsest congruence relation for A and every congruence relation for A is reflexive. Compare this with $\mathbb{I}(A)$, when $A=\langle A,=\rangle$. $\mathbb{I}(A)$ is in a sense "maximal" since $\mathbb{I}(A)=\mathbb{B}i(A)$. $\mathbb{K}(A)$ is in a sense "minimal" since $\mathbb{K}(A)$ only consists of the identity relation.

2.11   Lattices

In this section some well-known properties of lattices will be presented. We start off with the concepts of upper and lower bound.

Let $\langle X,\leq\rangle$ be a partially ordered set, $A\subseteq X$ and $x\in X$. The converse of $\leq$ is denoted by $\geq$.

x is an *upper bound* for A iff for all $a\in A$: $a\leq x$.

x is a *lower bound* for A iff for all $a\in A$: $x\leq a$.

x is a *least upper bound* for A iff x is an upper bound for A, and $x\leq y$ for
　　　all upper bounds y for A .

x is a *greatest lower bound* for A iff x is a lower bound for A and $x\geq y$
　　　for all lower bounds y for A.

If $\langle X,\leq\rangle$ is a partially ordered set and $A\subseteq X$, then there is at most one least upper bound for A and at most one greatest lower bound for A. Let

$$\text{lub}A = x \quad \& \quad \text{glb}A = y$$

mean that x is the least upper bound for A and y the greatest lower bound for A respectively.

A partially ordered set $\langle X,\leq\rangle$ such that every pair of elements in X has a least upper bound (in X) and a greatest lower bound (in X) is called a *lattice*. The least upper bound of $\{x,y\}$ is denoted by $x\vee y$ and the greatest lower bound of is denoted by $x\wedge y$. A partially ordered set $\langle X,\leq\rangle$ such that every subset of X has a least upper bound and a greatest lower bound is called a *complete lattice*. A complete lattice has a greatest element and a least element. If $\langle X,\leq\rangle$ is a partially ordered set such that every subset of X has a greatest lower bound, then $\langle X, \leq \rangle$ is a complete lattice. The least upper bound of a subset A is the greatest lower bound of the set of all upper bounds of A. (See for example Cohn, 1965, p. 21.)

Let $\langle X, \leq \rangle$ be a complete lattice. A subset C of X is a *closure system* in $\langle X, \leq \rangle$ if for every $D \subseteq C$, $glb\, D \in C$. (Note that $glb\, \varnothing$ is the greatest element in X, which thus also belongs to C.) If C is a closure system in $\langle X, \leq \rangle$, then $\langle C, \leq' \rangle$ is a complete lattice, where $\leq'$ is the restriction of $\leq$ to C. (See for example Birkhoff, 1967, p. 7.) Note that if $D \subseteq C$ then $lub\, D$ exists both in $\langle X, \leq \rangle$ and in $\langle C, \leq' \rangle$ but will in general be different, i.e. $lub\, D$ with respect to $\leq$ and with respect to $\leq'$ need not coincide.

PART TWO

AN INFORMAL PRESENTATION OF THE MAIN THEMES

CHAPTER 3

RELATIONALS

3.0 Introduction

As we pointed out in chapter 1, structural dependence is a relation between aspects, and aspects will be represented here by systems of relationals. In this chapter we will introduce the notion of a relational in detail. A relational is a representation of the ordinary language notion of relation and we shall use it in connections with aspects, rather than the usual set-theoretical representation of a relation as a set of ordered n-tuples. Formally a relational is a function which takes sets as arguments and sets of ordered n-tuples as values. The values of a relational are thus relations in the set-theoretical sense. The basic idea is that the value of the relational for a set is the extension on this set of the relation which the relational represents.

As the reader certainly has noticed, I use three notions of relations. I am aware of that I touch here upon controversial questions and probably substantial problems. However, I think they are of little importance for the aim of this study and I shall seek to avoid them. But let me try in spite of this to make somewhat clearer which view lies behind the last paragraph. The three notions of relations which are used are the following.

(1) The "prephilosophical" notion of relation which we use in ordinary language and in most sciences. To avoid confusion I shall sometimes use "relationship" when I mean relation in this sense.

(2) The notion of relation traditionally used in set theory, i.e. relation as a set of ordered n-tuples. A relation in this sense is often a kind of representation of a relationship.

(3) Relation as relational. A relational is another kind of representation of a relationship than a relation as a set of ordered n-tuples.

The representation of a relationship as a set of ordered n-tuples is in many contexts so common and immediate that we often forget that it is a matter of representation. We consider a relationship, for example "longer than", denote it by ρ and treat ρ as a set of ordered pairs and call both "longer than" and ρ for a "relation". But when we employ a different representation of a relationship than the usual one, I think it will be more obvious that there is a representation. I don't want to suggest that the character of this representation—someone might prefer to call it an analysis—is unproblematic, but it is not a problem for this investigation.

Relationships connected with aspects are what I will represent as relationals. In the sequel I talk interchangeably about things like "longer than" as
(a) a relationship (tacitly presupposing that it will be represented as a relational in the more formal reasoning)
(b) a relational (the relationship having thereby been represented)
(c) a relation in the ordinary set-theoretical sense (but this only in some special contexts where I discuss the usual treatment of "longer than").

Section 3.1 is devoted to a presentation of the basic conceptual framework of relationals. In section 3.2 we generalize properties of set-theoretical relations like symmetry and transitivity to relationals and introduce the important notions of the extension class and the characteristic class. Some examples are considered in section 3.3 and we introduce finite relational systems in section 3.4. We conclude with some historical and bibliographical remarks in section 3.5.

3.1   The fundamentals of relationals

An *n-ary relational* R is a function with sets as arguments and with sets of ordered n-tuples as values. The value of R for the argument A is a set of ordered n-tuples of elements in A. The domain of R, i.e. the set of arguments of R, will be called the *range of definition* of R and will be denoted $\mathbb{D}_R$ or just $\mathbb{D}$. The union of all sets in $\mathbb{D}$ is called the *universe of application* of R and will be denoted $\mathbb{U}_R$ or just $\mathbb{U}$. ($\mathbb{D}$ and $\mathbb{U}$ are used instead of $\mathbb{D}_R$ and $\mathbb{U}_R$, respectively, when there is no danger of confusion.) The range of definition is for many relationals the same as the set of non-empty subsets of the universe of application, but I will not suppose that this holds generally. Nor do I suppose that the universe of application is always an element in the range of definition, i.e. $\mathbb{U}_R \in \mathbb{D}_R$. However, this is the case in many applications. Note that in many contexts it is natural to think of $\mathbb{U}$, not as an actual totality, but as a "type", for example all physical objects, all rigid rods, all triangles or all distances.

The arguments of R will often be called domains for R. The set A is thus a *domain for* R iff $A \in \mathbb{D}_R$. Thus, the class of domains for R is the domain of the function R. To avoid confusion I will in the following not use "domain" in the second sense, but rather "range of definition". So the class of domains for R is the range of definition of R.

If $A \in \mathbb{D}$ then $R(A)$ is a set of ordered n-tuples of elements in A and is the *extension of* R *on* the domain A. Thus, the extensions of a relational are relations of the same arity. A relational is binary if its extensions are binary relations and, generally, a relational is n-ary if its extensions are n-ary. If $\langle x_1,...,x_n \rangle \in R(A)$, then R holds for $x_1,...,x_n$ (in that order) when $x_1,..,x_n$ are regarded as elements of A. $\langle x_1,...,x_n \rangle \in R(A)$ will often be written $R(A;x_1,..,x_n)$ or $R(A)(x_1,...,x_n)$, which can be read as "R holds between $x_1,...,x_n$ in A". If R is a binary relation, then it is often convenient to use $xR(A)y$ instead of $R(A)(x,y)$. An ordering relation will often be denoted by $\geq$, and $x \geq(A)y$ then means that $\geq$ holds between x and y in A. We will also sometimes write this as $x \geq y;A$.

A natural way to regard a relational is as a family of relations, where "relation" means a set of ordered n-tuples. The relational R is the family $\{R_A \mid A \in \mathbb{D}\}$ where R_A is the same as $R(A)$, i.e. the extension of R on A. The index set $\mathbb{D}$ of the family is the range of definition of the relational. Note also that an n-ary relational R can be regarded as a set of ordered n+1-tuples, i.e. $R = \{\langle A,x_1,..,x_n \rangle \mid R(A;x_1,..,x_n)\}$.

It will often be convenient to take the extension of R on A together with A, and I will therefore give the structure $\langle A,R(A) \rangle$ a special name, the *graph of* R *on* A, and I denote it $\underline{R}(A)$. Thus $\underline{R}(A)=\langle A,R(A) \rangle$. An interpretation is also given to $\underline{R}$ taken in isolation: $\underline{R}$ is the relational R when we let the values of R be the graphs rather than the extensions. $\underline{R}$ is thus just another form of R. The difference between graph and extension is small, and I shall in later chapters use the term "extension" even for the graphs and not distinguish between R and $\underline{R}$. It will be clear from the context what is intended.

The extension of a relation is often traditionally assumed to be the extension on the set of all objects the relation can be applied to. For relationals this corresponds to the definition that *the extension of* R *is* $R(\mathbb{U})$. One might be tempted to demand that the extension of R on A be the restriction of the extension of R to A, i.e. of $R(\mathbb{U})$ to A. This involves the tacit assumption that if $x_1,...,x_n$ belong to both

A and B, and A and B are members of $\mathbb{D}$, then

$$R(A;x_1,...,x_n) \text{ iff } R(B;x_1,...,x_n).$$

We shall, however, not assume that this holds generally, but call a relation for which it holds, *domain- stable*. A relation R is thus domain-stable if for all elements A and B in the range of definition it is the case that

$$R(A)/B = R(B)/A$$

(where $R(A)/B$ is the restriction of $R(A)$ to B, i.e. $R(A)/B=R(A)\cap B^n$; for further details see section 2.1). I have chosen the term "domain-stable" for the following reason. If R has this property, it holds or does not hold between $x_1,...,x_n$ independently of which domain we consider, i.e. independently of the domain (containing $x_1,...,x_n$) on which we view the extension of R. The truth-value of the proposition that R holds between $x_1,...,x_n$ is therefore stable with regard to change of the domain as far as $x_1,...,x_n$ belong to the altered domain.

There is a condition on R related to domain stability which will play a certain role later on. This condition will be called *weak domain stability* and R satisfies it iff

$$R(B) = R(A)/B$$

for A,B in $\mathbb{D}$ such that $B \subseteq A$. It is easy to see that if R is domain-stable then R is weakly domain-stable. And if $\mathbb{D}$ is closed under intersection, i.e. for all A,B in $\mathbb{D}$ $A \cap B$ belongs to $\mathbb{D}$, then the converse holds, too. (See section 7.1 for proof.)

The word "stability" is used in several senses in logic and mathematics, which motivates the qualification "domain". However, within the confines of this book the risk of confusing domain stability with other notions of stability is small and the attribute "domain" will usually be dropped and we will be content with the more simple term *stability*. The term "weak domain stability" will be correspondingly simplified to *weak stability*.

It is obvious from the definition of weak stability that if U_R belongs to $\mathbb{D}_R$ and R is weakly stable, then for all A in $\mathbb{D}_R$, $R(A)=R(U_R)/A$. $R(U_R)$ is a set of ordered n-tuples, so in this special case all information about R can be given by the range of definition $\mathbb{D}$ and a set of ordered n-tuples since the values of R are restrictions of this set to the elements in $\mathbb{D}$. The difference in this case between representing an n-ary relationship as a set of ordered n-tuples and as a relational may seem subtle (but is not unimportant, as we shall see). The information brought about by a relational which is not stable cannot, however, be so easily expressed by a set of ordered n-tuples. One of the reasons for introducing relationals is that many relationships in science and ordinary life are not well represented by sets of ordered n-tuples but by relationals that are not stable.

An extreme example is the unary relationship "being best" in a certain respect, for example in mathematics. The best in the class is not necessarily the best in the school, and being best in the school is of course not necessarily being best in the country. This is so obvious that we almost always explicitly mention the domain we consider or understand it tacitly. And I think that we usually regard "being best in mathematics" as elliptical and therefore refuse to represent it as a set-theoretical relation, but it can suitably be represented as a relational—a highly unstable one.

The relational "being best" (in a certain respect) is thus of its nature unstable. That is not the case with the relational "at least as good as" (in the same respect). But it is a well-known fact from psychology that a person's preference between x and y often depends on which other alternatives the person has to choose between. If a person compares x and y he can find them equally good. But if he compares x,y and z he may find that x is better than z but that y and z are equally good and hence conclude that x is in fact better than y. Whether x and y are judged equally good can therefore depend on whether there is an element z or not in the domain under consideration such that one of x and y is judged equal to z but not to the other. (We are here touching on the subject of semiorders to which we shall return in section 3.3.) But also when the judgement is made on the basis of quite different aspects the comparison between x and y can depend on the presence of other alternatives. Preference comparisons are thus rather often dependent on the actual domain, and that is especially true of the equality part, i.e. the relation "as good as". But it is here a matter of dependence of a more "accidental" character than with "being best". It is, for example, rather common and not obviously elliptic to omit reference to the actual domain and assert that x is at least as good as y. So preference relationals are often unstable, but the instability is of another character than for the relational "being best".

The notion of a relational can formally be regarded as a kind of generalization of the notion of relation in the sense of a set of ordered n-tuples; a relational is essentially a family of relations. Operations, attributes etc. which are applicable to relations can therefore often be redefined in such a way that they become applicable to relationals. We have already seen one example, namely arity, and we shall study several examples of this in the next section. Let us for the moment take a look at just one more simple example: the conversion operation. If ρ is a binary relation then the converse of ρ, denoted ρ^c, is the set

$$\{\langle x,y \rangle \mid y\rho x\} \ .$$

Let us define the converse of the relational R as the relational R^c such that for all $A \in \mathbb{D}_R$

$$R^c(A) = R(A)^c \ .$$

And analogously we define $\underline{R}^c$ by

$$\underline{R}^c(A) = \langle A, R^c(A) \rangle \ .$$

The converse of a relational is thus the converse of all of its extensions.

A relational is a function, which implies that operations, attributes etc. that are applicable to functions are also applicable to relationals. It is therefore meaningful to ask if a relational R is a one-to-one function. If we take the values of the relationals to be the graphs and not the extensions and thus consider $\underline{R}$ instead of R, then we always get a one-to-one function, which is easily proved: If $A \neq B$ then obviously $\langle A,R(A) \rangle \neq \langle B,R(B) \rangle$, since two structures must have the same domain to be equal. It is therefore always meaningful to construct the inverse of R, which we, in accordance with tradition, denote $\underline{R}^{-1}$. And if the inverse of R exists we denote it of course by R^{-1}. It is important to distinguish $\underline{R}^{-1}$ from $\underline{R}^c$ and R^{-1} from R^c, i.e. distinguish the inverse of a relational from the converse of a relational. The inverse of a relational R is the inverse of R considered as a function, and it is usually not a relational. The converse of R is the application to R of the conversion operation for relations (as sets of ordered n-tuples) generalized to relationals in such a way that we take the converse of all the extensions of R. It is of course also possible to take the conversion operation and apply it to R considered as a function and take the inversion operation and generalize it to R by constructing the inverses of all the extensions of R (if they exist). The choice I have made for the definitions of converse and inverse for relationals is dictated by practical considerations, and we shall return to this later on. What is of interest just now is to point out that when we come to take notions from the theory of relations and functions as we find it in mathematics, we have a choice between two ways of applying them to relationals, either to a relational as a function or to each and every extension of the relational. In most cases , however, the choice is very simple; for a given notion there is often only one way of applying it to relationals which is of interest. Sometimes there are two or more different terms used for (almost) the same thing, and we shall then use one term for relationals regarded as functions and another for relationals as generalizations of relations when both are of interest. As an example consider relative product and functional composition. Relative product $|$ will be used in the latter way, i.e. the relative product of the binary relationals R and S with the same range of definition is denoted $R \, | \, S$ and defined by

$$(R \, | \, S) \, (A) \; = \; R(A) \, | \, S(A)$$

The relative product of two binary relationals with the same range of definition $\mathbb{D}$ is thus a relational with $\mathbb{D}$ as range of definition. The composition of two relationals R and S is denoted $R \, o \, S$ and defined according to the usual definition of composition.

A relational R_0 representing "being best" (in some respect) is, as we have already seen, hardly weakly stable. But we can split the condition of weak stability into two halves and it seems reasonable that R_0 satisfies one of these halves. Let us take a look at this in detail.

We say that R satisfies the condition of *upward stability* if for all $A, B \in \mathbb{D}$,

$$B \subseteq A \quad \text{implies} \quad R(B) \subseteq R(A) \, .$$

And we say that R satisfies the condition of *downward stability* if for all $A, B \in \mathbb{D}$,

$$B \subseteq A \quad \text{implies} \quad R(B) \supseteq R(A)/B \, .$$

Note that if $R(B) \subseteq R(A)$ and $B \subseteq A$ then $R(B) \subseteq R(A)/B$, since $R(B) \subseteq B^n$ and hence $R(B) \subseteq R(A) \cap B^n$. (We presuppose that R is n-ary.) This means that R is upward stable if for all $A, B \in \mathbb{D}$,

$$B \subseteq A \quad \text{implies} \quad R(B) \subseteq R(A)/B \, .$$

Then it is obvious that R is both upward and downward stable iff R is weakly stable. Let us say that R satisfies the condition of *directed stability* if R is upward or downward stable.

It may be clarifying to note that R is upward stable iff for all A,B in $\mathbb{D}$ such that $B \subseteq A$ it holds that $R(B; x_1, ..., x_n)$ implies $R(A; x_1, ..., x_n)$ for all $x_1, ..., x_n$ in B. And R is downward stable iff for all $A, B \in \mathbb{D}$ such that $B \subseteq A$ it holds that $R(A; x_1, ..., x_n)$ implies $R(B; x_1, ..., x_n)$ for all $x_1, ..., x_n$ in B.

Let us now return to R_0, i.e. "being best" (in some respect). It is at least reasonable to ask if R_0 is downward stable. Suppose the following holds: If x is a member of B, x is one of the members of A which are best in the respect at issue and $B \subseteq A$, then x is one of the members of B which are best. Then the relational R_0 which represents "being best" in the respect at issue is downward stable, i.e. if $B \subseteq A$ and $x \in B$ then $R_0(A; x)$ implies $R_0(B; x)$.

The reader familiar with social choice theory will probably make an association with Sen's property α of choice functions at this point. Property α states, in Sens's words, "that if some element of subset S_1 of S_2 is best in S_2, then it is best in S_1." (See Sen ,1970, p. 17.) Formally property α has the following wording:

$$x \in S_1 \subset S_2 \rightarrow [x \in C(S_2) \rightarrow x \in C(S_1)] \text{, for all } x .$$

(See Sen, 1970, p. 17.)

The formal similarity between downward stability and Sen's property α seems striking. It may give rise to the supposition of a close connection between choice functions and one kind of relationals. According to Sen property α is "a very basic requirement of rational choice, and in a different context has been called the condition of 'the independence of irrelevant alternatives' ". We will take a closer look at the use of relationals in social choice theory and the connection between stability and "independence of irrelevant alternatives" in chapter 5.

Let me end this section with a remark on the elusive question of what a relational really is (its true nature). Relationals are intended to function as representations of relations in ordinary language and thereby offer an alternative to set-theoretical relations as sets of ordered n-tuples. Now it is natural to ask whether relationals themselves are sets and thus set theoretical objects. It is obvious that

$$\{\langle A, R(A) \rangle \,|\, A \in \mathbb{D}\}$$

is a set-theoretical object but it is not obvious that this set and R are the same thing. R is a function but a function need not be understood as a set. An alternative way of regarding functions is as "rules" or "programs", and I think this is not seldom done in mathematics, especially nowadays under the influence of computer science. I like to think of relationals as "rules" or "programs", but it is not at all necessary for what is said in this book that relationals be understood in this way. It is the logical (or mathematical) form of relationals which is essential here, not their "true nature".

3.2   Formal properties of relationals

There are lots of well-known formal properties for set-theoretical relations, for example transitivity, symmetry, reflexivity and connectedness (see section 2.2). We want of course to be able to apply these properties even to relationals. Let us say that a binary relational R is transitive (symmetric, reflexive, etc.) if all the values of R are transitive (symmetric, reflexive etc.). We proceed in the same way with more complicated set-theoretical properties of relations, for example partial order, linear order and weak order. In an attempt to minimize the risk of confusion a distinction is drawn between "ordering" and "order"; an ordering is a relational while an order is a relation in the usual set-theoretical sense. A binary relational R is for example a weak ordering if for all elements A in the range of definition $\mathbb{D}$, R(A) is a weak order. Hence, the binary relational R is a weak ordering iff for all A in the range of definition $\mathbb{D}$ and all x,y,z in A the following two conditions are satisfied:

(i) Either R(A;x,y) or R(A;y,x).

(ii) If R(A;x,y) and R(A;y,z) then R(A;x,z).

Generally, we extend the application of set-theoretical predicates to relationals by the following rule: R is a $\mathbb{P}$ iff every graph of R is a $\mathbb{P}$. Instead of "R is a $\mathbb{P}$" we shall often say "R satisfies $\mathbb{P}$".

Define $\mathbb{E}_R$ as the set $\{\underline{R}(A) \mid A \in \mathbb{D}\}$, and call it the *extension class* of R. Thus the extension class of R is the set consisting of all the graphs of R. However, it is sometimes more convenient to let $\mathbb{E}_R$ consist of the extensions of R and we will occasionally do so, often without explicitly mentioning it—it is usually quite clear from the context what is intended. In later chapters we will disregard the difference between graph and extension almost completely. If R(A)=$\langle A, \rho \rangle$ we shall say that both $\langle A, \rho \rangle$ and ρ are the extensions of R on A. In the few cases where it is necessary to emphasize the distinction between $\langle A, \rho \rangle$ and ρ we will call ρ the *proper extension* of R on A and $\langle A, \rho \rangle$ the graph of R on A. The reason why this is mentioned now is only to justify the term "extension class". Note that $\mathbb{E}_R$ is the range of R regarded as a function.

Let the *characteristic class* of R be $\mathbb{E}_R$ closed under isomorphisms and denote it by $\mathbb{C}_R$. Thus

$$\mathbb{C}_R = \{X \mid \exists A \in \mathbb{D}: \underline{R}(A) \cong X\}$$

where $\cong$ means "isomorphic to".

$\mathbb{C}_R$ is a class of relational structures of the same type closed under isomorphisms; i.e. $\mathbb{C}_R$ is a theory given by an extrinsical characterization (see Scott and Suppes, 1958, and Suppes 1967, p.1-9). It is of course often possible to characterize $\mathbb{C}_R$ intrinsically, i.e. as a set of sentences or, preferably, by an axiom system. A theory $\mathbb{T}$ which has $\mathbb{C}_R$ as its class of models is [a formulation of] the *extension theory* of R. If for all elements A in $\mathbb{D}$, $\underline{R}(A)$ is a model of a theory $\mathbb{T}$, then we say that R satisfies $\mathbb{T}$. A relational satisfies of course its extension theory. But a theory $\mathbb{T}$ which it satisfies is not necessarily its extension theory since every model of $\mathbb{T}$ need not belong to $\mathbb{C}_R$.

It is customary in measurement theory to avoid dealing in detail with the language formulating measurement systems such as difference systems, extensive systems and so on. We will follow this practice here. Theories will therefore often be characterized by set-theoretical predicates. Thus, the set-theoretical predicate $\mathbb{P}$ *formulates* (or *is*) the *extension theory* of R if $\mathbb{C}_R$ is the set of structures that satisfy $\mathbb{P}$, i.e. for every A in $\mathbb{D}$, R(A) is a $\mathbb{P}$ and for every $\mathbf{Y}$ which is a $\mathbb{P}$ there is an element A in $\mathbb{D}$ such that $\underline{R}(A)$ is isomorphic to $\mathbf{Y}$.

Two set-theoretical predicates $\mathbb{P}$ and $\mathbb{P}'$ will be called *extensionally equivalent* if it is the case that for any $\mathbf{X}$

$$\mathbf{X} \text{ is a } \mathbb{P} \quad \text{iff} \quad \mathbf{X} \text{ is a } \mathbb{P}'.$$

Thus, if both $\mathbb{P}$ and $\mathbb{P}'$ are (or formulate) the extension theory of R, then $\mathbb{P}$ and $\mathbb{P}'$ are extensionally equivalent.

It is important to note the difference between the following two conditions on a binary relation R.

(1) R is transitive, i.e. for all domains A in $\mathbb{D}$ and all x,y,z in A: R(A;x,y) and R(A;y,z) implies R(A;x,z)

(2) for all x,y and z in $\mathbb{U}$ such that $\{x,y\}$, $\{y,z\}$ and $\{x,z\}$ in $\mathbb{D}$: $R(\{x,y\};x,y)$ and $R(\{y,z\};y,z)$ implies $R(\{x,z\};x,z)$.

If (2) is satisfied, then the relational R will be said to be *pointwise transitive*. The "strength" of pointwise transitivity depends of course on which "points" belong to $\mathbb{D}$. Let us say that the n-ary relational R is *local* iff $\{x_1,...,x_n\}$ in $\mathbb{D}$ for all $x_1,...,x_n$ in $\mathbb{U}$. This means that an n-ary relational is local iff every non-empty set of n elements or fewer taken from $\mathbb{U}$ belongs to $\mathbb{D}$. If R is local we can define a new relational R* with $\mathbb{D}$ as its range of definition in the following way:

$$R^*(A) = \{\langle x_1,...,x_n \rangle \mid x_1,...,x_n \in A \ \& \ R(\{x_1,...,x_n\};x_1,...,x_n)\} \ .$$

As we shall see in the next section, R*(A) is sometimes a kind of "raw data" for the extension of R on A, and this happens when R(A) is a refinement or transformation of R*(A). R* can then be regarded as providing the raw data for R. It is easy to see that R* is domain-stable and it holds for every A in $\mathbb{D}$ and all $x_1,...,x_n$ in A that

$$R^*(A;x_1,...,x_n) \quad \text{iff} \quad R(\{x_1,...,x_n\};x_1,...,x_n) \tag{3.2.1}$$

Note that if R is local and stable, then R equals R*. This follows since the locality and stability of R implies for all A in $\mathbb{D}$ and all $x_1,...,x_n$ in A

$$R(A;x_1,...,x_n) \quad \text{iff} \quad R(\{x_1,...,x_n\};x_1,...,x_n)$$

and from (3.2.1) then follows

$$R(A;x_1,...,x_n) \quad \text{iff} \quad R*(A;x_1,...,x_n) .$$

Note also that it does not generally hold that $R*(\{x_1,...,x_n\}) = R(\{x_1,...,x_n\})$. To see this, let us suppose that R is a ternary relational. Suppose further that R is local and that x,y,z are distinct elements of $\mathbb{U}$. According to the definition of R*

$$R*(\{x,y,z\};x,x,y) \quad \text{iff} \quad R(\{x,y\};x,x,y) .$$

But $R(\{x,y\};x,x,y)$ is not necessarily equivalent to $R(\{x,y,z\};x,x,y)$, so $R*(\{x,y,z\};x,x,y)$ is not necessarily equivalent to $R(\{x,y,z\};x,x,y)$.

Let us introduce a concept which is related to pointwise transitivity. We say that R is *regionally transitive* if R is local and R* is transitive. If R is local and pointwise transitive then R is regionally transitive. The converse does not hold. We can have $R(\{x,y\};x,y)$, $R(\{y,z\};y,z)$ and $\neg R(\{x,z\};x,z)$, but if there is no domain A in $\mathbb{D}$ which contains x,y and z then R* can still be transitive. If R is local and for all x, y and z in $\mathbb{U}$ there is a domain A in $\mathbb{D}$ such that $\{x,y,z\}$ in A, then R is regionally transitive iff R is pointwise transitive.

If R is regionally transitive it does not follow that R is transitive. And if R is local and transitive it does not follow that R is regionally transitive. We shall return to the last fact in the next section.

It is obviously possible to define a "regional" version of an arbitrary set-theoretical predicate $\mathbb{P}$ according to the rule

$$R \text{ is a regional } \mathbb{P} \quad \text{iff} \quad R \text{ is local and } R* \text{ is a } \mathbb{P}.$$

We can for example talk about regional equivalence relationals and regional weak orderings. A regional equivalence relational is of course a relational which is regionally transitive, regionally symmetric and regionally reflexive. Note that a local relational which is symmetric is also regionally symmetric. The same holds for reflexivity. However, if R is regionally reflexive it does not follow that R is reflexive. But it is not easy to find an empirically interesting example of a relational which is regionally reflexive but not reflexive. Analogously, if R is regionally symmetric it does not follow that R is symmetric. And to continue the analogy, it seems not to be the case that science is teeming with interesting relationals which are regionally symmetric but not symmetric.

We shall make use of regional properties in the next section, where we study some examples.

No name has yet been given to R*, and we shall now remedy this situation. Let us say that R* is the *regionalization* of R since R* is in a rough sense "what holds locally for R". R is thus a regional $\mathbb{P}$ iff R is local and the regionalization of R is a $\mathbb{P}$.

R* has been defined only when R is local. However, R* can also be defined when a weaker condition on R is satisfied. Let us say that the n-ary relational R is *weakly local* if $\{x_1,...,x_n\}$ in $\mathbb{D}$ for all $x_1,...,x_n$ in U such that there is A in $\mathbb{D}$ containing $x_1,...,x_n$ as elements. If R is weakly local, then R* can obviously be defined. R is said to be a weakly regional $\mathbb{P}$ iff R is weakly local and R* is a $\mathbb{P}$.

Note that if R is an n-ary relational with $\mathbb{D}$ as range of definition and R is local, then every n-ary relational S with $\mathbb{D}$ as range of definition is local too. The property of being local is thus a property of the range of definition of the relational. Let us say, then, that a set M of non-empty sets is *n-ary local* if $\{x_1,...,x_n\}$ in M for all $x_1,...,x_n$ in $\cup$M (where $\cup$M is the union of all sets in M). Weak locality of M is defined analogously.

3.3 Some examples

Which formal properties a relation "greater than or equal to" (in a certain respect) has is a much discussed question in the measurement-theoretic literature. It is of course an order, but of which kind? It is usually accepted that "greater than" is irreflexive, asymmetric and transitive while "equal to" is reflexive and symmetric. The controversial point is whether "equal to" is a transitive relation. For some authors it is in many cases definitely so. They mean that "greater than or equal to" is a weak order, and thus that "greater than" is a strict weak order and "equal to" is an equivalence relation. For others this is highly questionable. They suggest that in many cases the relation "greater than" is not a strict weak order but rather a semi-order, and consequently equality is not an equivalence relation. (Note that if the set-theoretical relation ρ is a strict weak order, then the corresponding indifference relation—i.e. σ defined by $x\sigma y$ iff $\neg x\rho y$ and $\neg y\rho x$ —is an equivalence relation. But if π is a semiorder, then the corresponding indifference relation σ defined analogously by $x\sigma y$ iff $\neg x\pi y$ and $\neg y\pi x$ is not necessarily an equivalence relation. See section 2.3 and 2.5)

Turning to relationals, it seems to me that it is not unreasonable to say that the controversy between weak ordering and semiordering could be resolved in some contexts (but certainly not all) by saying that equality is an equivalence relational but not always a regional equivalence relational. And that "greater than" is a strict weak ordering, but not always a regional strict weak ordering. On the other hand "greater than" is a regional semiordering. This point is perhaps worth elaborating. (Basic definitions and elementary results about semiorders are presented in section 2.5.)

Suppose S is a stable local semiordering, so that S(A) is a semiorder for all A in $\mathbb{D}$. As is well-known (see section 2.5 for references), if $\langle X, \zeta \rangle$ is a semiorder, then $\langle X, \rho \rangle$, where ρ is defined by

$$x\rho y \text{ iff } \forall z \in X: z\zeta x \Rightarrow z\zeta y \text{ \& } y\zeta z \Rightarrow x\zeta z,$$

is a weak order. Therefore, if we take R to be such that for all A in $\mathbb{D}$

$$R(A) = \{\langle x,y \rangle \in A^2 \mid \forall z \in A: (S(A;z,x) \Rightarrow S(A;z,y)) \text{ \& } S(A;y,z) \Rightarrow S(A;x,z))\}$$

then R is a weak ordering. We call R the *weak ordering associated with* S. Define P as the strict part of R, i.e. for all A in $\mathbb{D}$ and all x and y in A

$$P(A;x,y) \text{ iff } R(A;x,y) \text{ and } \neg R(A;y,x).$$

We call P the *strict weak ordering associated with* S. P(A) is thus a strict weak order for all A in $\mathbb{D}$, and so P is a strict weak ordering. Since S is local and $\mathbb{D}_R = \mathbb{D}_P = \mathbb{D}_S = \mathbb{D}$, R and P are local. Note that

$$S(\{x,y\};x,y) \text{ iff } (R(\{x,y\};x,y) \text{ \& } \neg R(\{x,y\};y,x))$$

and hence

$$S(\{x,y\};x,y) \text{ iff } P(\{x,y\};x,y). \tag{3.3.1}$$

Since P is local, P* is defined, and for all A in $\mathbb{D}$ and x,y in A,

$$P^*(A;x,y) \text{ iff } P(\{x,y\},x,y). \tag{3.3.2}$$

And since S is stable,

$$S(A;x,y) \text{ iff } S(\{x,y\};x,y) \tag{3.3.3}$$

for all A in $\mathbb{D}$ and x,y in A. (3.3.1), (3.3.2) and (3.3.3) together give

$$P^*(A;x,y) \text{ iff } S(A;x,y) \tag{3.3.4}$$

for all A in $\mathbb{D}$ and x,y in A. Hence, P* equals S. P is thus a regional semiordering, and P is at the same time a strict weak ordering.

Let Q be the indifference part of R, i.e. Q is a relational with $\mathbb{D}$ as range of definition defined by

$$Q(A;x,y) \text{ iff } R(A;x,y) \text{ \& } R(A;y,x).$$

Hence

$$Q(A;x,y) \text{ iff } \neg P(A;x,y) \text{ \& } \neg P(A;y,x).$$

Q is of course an equivalence relational. We call Q the *equivalence relational associated with* S.

Let J be the indifference part of S, i.e. J has $\mathbb{D}$ as its range of definition and J is defined by

$$J(A;x,y) \text{ iff } \neg S(A;x,y) \And \neg S(A;y,x).$$

Since S is stable this holds for J, too. Q and J are of course local. Q* therefore exists and the definitions of Q*, Q, P* and J, together with (3.3.1), (3.3.4) and the stability of J, give the following series of equivalences:

$$Q^*(A;x,y) \text{ iff } Q(\{x,y\};x,y) \text{ iff } \neg P(\{x,y\},x,y) \And \neg P(\{x,y\};y,x) \text{ iff}$$
$$\text{iff } \neg P^*(\{x,y\},x,y) \And \neg P^*(\{x,y\};y,x) \text{ iff } \neg S(\{x,y\},x,y) \And \neg S(\{x,y\};y,x) \text{ iff}$$
$$J(\{x,y\};x,y) \text{ iff } J(A;x,y).$$

Hence, Q* equals J.

The equivalence relation associated with a semiorder can, as was pointed out in section 2.5, be defined in two ways. The same holds for equivalence relationals. We can therefore define Q by

$$Q(A;x,y) \text{ iff } \forall z \in A: J(A;x,z) \Leftrightarrow J(A;y,z)$$

or, since $J(A;x,y)$ iff $Q^*(A;x,y)$, by

$$Q(A;x,y) \text{ iff } \forall z \in A: Q^*(A;x,z) \Leftrightarrow Q^*(A;y,z) \qquad (3.3.5)$$

This construction of a weak order R, of a strict weak order P and of an equivalence relational Q out of the stable, local semiordering S holds for all such orderings. We can therefore choose S such that it is not a strict weak ordering. Then J is not an equivalence relational, since it is not transitive. We then have the situation that P is a strict weak ordering and a regional semiordering, but that P is not a regional strict weak ordering. Furthermore, Q is an equivalence relational, but not a regional equivalence relational, since it is not regional transitive.

Since we have supposed that S is stable, it follows that J is stable, too. But neither R nor P nor Q is necessarily stable. To see this, suppose that

$$S(\{x,y,z\}) = \{\langle x,z \rangle\}. \qquad (3.3.6)$$

If we interpret S as "definitely greater than", then (3.3.6) means that x is definitely greater than z, but none of x and y is definitely greater than the other and none of y and z is definitely greater than the other. From the construction of R,P and Q follows

$$R(\{x,y,z\}) = \{ \langle x,y \rangle, \langle x,z \rangle, \langle y,z \rangle, \langle x,x \rangle, \langle y,y \rangle, \langle z,z \rangle \}$$
$$P(\{x,y,z\}) = \{ \langle x,y \rangle, \langle x,z \rangle, \langle y,z \rangle \}$$
$$Q(\{x,y,z\}) = \{ \langle x,x \rangle, \langle y,y \rangle, \langle z,z \rangle \}.$$

Since S is stable

$$S(\{x,y\}) = \varnothing.$$

Hence,

$$R(\{x,y\}) = \{ \langle x,y \rangle, \langle y,x \rangle, \langle x,x \rangle, \langle y,y \rangle \}$$
$$P(\{x,y\}) = \varnothing$$
$$Q(\{x,y,z\}) = \{ \langle x,y \rangle, \langle y,x \rangle, \langle x,x \rangle, \langle y,y \rangle \} .$$

$R(\{x,y,z\})/\{x,y\} \neq R(\{x,y\})$, and the situation is analogous for P and Q. (In Danielsson, 1974, p. 32f. the idea of—what is here called—stability is applied to weak orders associated with semiorders, and the discussion above is inspired by this.)

R,P and Q are thus unstable. However, P is upward stable while R and Q are downward stable. To see this let $A,B \in \mathbb{D}$ and $B \subseteq A$. Since S is stable

$$S(A)/B = S(B).$$

$R(A)$ and $R(B)$ are the weak orders associated with $S(A)$ and $S(B)$, respectively. $P(A)$ and $P(B)$ are the strict weak orders respectively associated with $S(A)$ and $S(B)$. Finally, $Q(A)$ and $Q(B)$ are the equivalence relations associated with $S(A)$ and $S(B)$. Since $\langle A, S(A) \rangle$ and $\langle B, S(B) \rangle$ are semiorders such that $B \subseteq A$ and $S(B) = S(A)/B$ we can apply (2.5.2). Hence,

$$P(B) \subseteq P(A) \ \& \ R(B) \supseteq R(A)/B \ \& \ Q(B) \supseteq Q(A)/B.$$

This shows that P is upward stable while R and Q are downward stable.

We have seen that there are relationals which are both strict weak orderings and regional semiorderings but not regional strict weak orderings. One might therefore wonder whether there are any necessary connections between the properties of a relational R and the properties of the regionalization R* of R. This question is related to the problem of to what extent R and R* are dependent on (determined by) each other, which will be discussed in section 4.4. Note that for P and P* as constructed above, P is completely determined by P* (see further section 4.4), and still P and P* can have rather different formal properties. The general question how much the formal properties of a relational R and its regionalization R* can differ when R is completely determined by R* might perhaps be an interesting question but it is not pursued here.

In section 3.2 it was maintained that R* may in some cases be regarded as providing the raw data for R. To justify this, let us look at an example.

Let N be a given individual whose ability to judge loudness we want to investigate. Let P be the relational "louder than according to N" and Q the relational "equally loud as according to N". It does not seem unreasonable to use pairwise comparisons to determine N's judgement of loudness. We therefore let N compare pairs of sounds x and y and decide if x is louder than y or if y is louder than x or if x and y are equally loud. (We suppose that exactly one of these judgements hold.) If N judges x to be louder than y we take this to mean that $P(\{x,y\};x,y)$. If N judges y to be louder than x we analogously interpret this as $P(\{x,y\};y,x)$. Finally, if N judges x and y to be equally loud we interpret this as $Q(\{x,y\};x,y)$. Assuming that P and Q are local, which is natural in connection with pairwise comparisons, we can construct P^* and Q^*. $P^*(A)$ may now be regarded as containing the basic information for determining $P(A)$. With $P^*(A)$ as the basic information about the elements of A with respect to loudness we construct $P(A)$ and $Q(A)$. $P^*(A)$ may therefore be regarded as the raw data for the data on loudness of the elements in A according to N. Let us for simplicity suppose that $P^*(A)$ is a semiorder and that this holds for all A in the range of definition of P. Hence, P^* is a semiordering and P a regional semiordering. It may then be natural to let $P(A)$ be determined as the strict weak order associated with $P^*(A)$, and thus to let P be the strict weak ordering associated with P^*. P is then completely determined by P^* (in the sense specified in chapter 4). In this case P^* may be called the raw data giver for P. There is a natural interpretation of P^* in empirical terms: "definitely louder than according to N". If $P^*(A;x,y)$ then $P^*(B;x,y)$ for all B in the range of definition such that x and y in B.

If P is the strict weak ordering associated with P^*, then a natural thing to do is to let Q be the equivalence relational associated with P^*, in which case it holds that
$$Q(A;x,y) \text{ iff } \neg P(A;x,y) \,\&\, \neg P(A;y,x) \,.$$
Q^* functions then as the raw data giver for Q since it holds that
$$Q(A;x,y) \text{ iff } (\forall z \in A: Q^*(A;x,z) \Leftrightarrow Q^*(A;y,z)).$$
(see 3.3.5 which is applicable here). In terms of the explication of complete determination stated in the next chapter, Q is completely determined by Q^*. It is also the case that Q is completely determined by P^* but P is not completely determined by Q^*.

We now turn to another example. It will illustrate how we can get relationals in a simple and natural way whose formal properties differ considerably from those of their regionalizations. (The basic idea of this example is due to Rabinowicz, 1989.)

Let P_1 and P_2 be strict weak orderings with the same range of definition $\mathbb{D}$. Further, let P_1 and P_2 be local and P_1 a regional semiordering. We assume that P_1 is the strict weak ordering associated with P_1^* and that P_2 is stable. Thus, P_2 is a regional strict weak ordering. The equivalence relationals corresponding to P_1 and P_2 will be denoted Q_1 and Q_2, respectively. Let P_0 be the lexicographic ordering relative to P_1 and P_2. Then P_0 has $\mathbb{D}$ as its range of definition and for all A in $\mathbb{D}$ and x,y in A

$$P_0(A;x,y) \Leftrightarrow P_1(A;x,y) \vee (Q_1;A;x,y) \,\&\, P_2(A;x,y)) \,.$$

From elementary results on lexicographic orders it follows that P_0 is a strict weak ordering. Let Q_0 be the equivalence relational corresponding to P_0, i.e.

$$Q_0(A;x,y) \Leftrightarrow \neg P_0(A;x,y) \,\&\, \neg P_0(A;y,x) \,.$$

Let us now ponder which regional properties P_0 and Q_0 have. It does not hold generally that P_0 is regionally transitive. To see this suppose that x,y and z are distinct elements in the range of application $\mathbb{U}$ of P_0 and that $\{x,y,z\}$ belongs to $\mathbb{D}$. Since P_0 is local $\{x,y\}$, $\{y,z\}$ and $\{x,z\}$ are elements in $\mathbb{D}$. Suppose further that

$$Q_1(\{x,y\};x,y) \,\&\, Q_1(\{y,z\};y,z) \,\&\, P_1(\{x,z\};z,x)$$
$$P_2(\{x,y\};x,y) \,\&\, P_2(\{y,z\};y,z) \,\&\, P_2(\{x,z\};x,z) \,.$$

This is compatible with the formal properties we have assumed for P_1, Q_1, P_2 and Q_2; P_1 is a regional semiordering and P_2 a regional strict weak ordering. According to the definition of P_0 we get

$$P_0(\{x,y\};x,y) \,\&\, P_0(\{y,z\}y,z) \,\&\, P_0(\{x,z\};z,x)$$

and hence

$$P_0{}^*(\{x,y,z\}) = \{\ \langle x,y\rangle\,,\langle y,z\rangle\,,\langle z,x\rangle\ \} \,.$$

P_0 is thus not regional transitive and is therefore not a regional semiordering. Since P_1 is the strict weak ordering associated with $P_1{}^*$ it holds that

$$P_1(\{x,y,z\}) = \{\ \langle z,y\rangle\,,\langle y,x\rangle\,,\langle z,x\rangle\ \}$$

Hence,

$$P_0(\{x,y,z\}) = \{\ \langle z,y\rangle\,,\langle y,x\rangle\,,\langle z,x\rangle\ \} \,,$$

so $P_0(\{x,y,z\};y,x)$. Since $P_0(\{x,y\};x,y)$, P_0 is neither upward nor downward stable. However, P_0 satisfies a certain kind of restricted or partial stability. The reason is that P_1 is upward stable and Q_1 is downward stable and P_0 depends to a great extent on P_1 and Q_1. In fact, if $B \subseteq A$ and $P_1(B;x,y)$ then $P_0(A;x,y)$. Or, expressed in another way, if $\neg Q_1(B;x,y)$ then $B \subseteq A$ and $P_0(B;x,y)$ (together) implies $P_0(A;x,y)$. The following definitions are an attempt to catch the ideas behind partial or restricted stability.

Let us say that

R is *upward stable for* $\langle x_1,...,x_n\rangle$ *from* A if $x_1,...,x_n$ in A and for all B,C in $\mathbb{D}_R$ such that $B,C \supseteq A$, $B \subseteq C$ implies

$$R(B;x_1,...,x_n) \Rightarrow R(C;x_1,...,x_n).$$

R is *downward stable for* $\langle x_1,...,x_n\rangle$ *from* A if $x_1,...,x_n$ in A and for all B,C in $\mathbb{D}_R$

such that $B,C \subseteq A$ and $x_1,...,x_n$ belong to B, $B \subseteq C$ implies

$$R(C;x_1,...,x_n) \Rightarrow R(B;x_1,...,x_n) .$$

It is easy to see that R is upward stable for $\langle x_1,...,x_n \rangle$ from A for all A in $\mathbb{D}_R$ and all $x_1,...,x_n$ in A iff R is upward stable. And R is downward stable for $\langle x_1,...,x_n \rangle$ from A for all A in $\mathbb{D}_R$ and all $x_1,...,x_n$ in A iff R is downward stable.

It is natural to say that R is upward stable for $\langle x_1,...,x_n \rangle$ *starting from* A if R is upward stable for $\langle x_1,...,x_n \rangle$ in A and for all B such that $B \subset A$, R is not upward stable for $x_1,...,x_n$ from B. And analogously for downward stability.

The notions of partial or restricted stability may be of particular interest as tools for analyzing situations involving decisions.

Let us now return to P_0. P_0 is upward stable for $\langle x,y \rangle$ from A for all A in $\mathbb{D}$ and all x,y in A such that $\neg Q_1(A;x,y)$. This can be shown in the following way. Suppose x,y are in A and $\neg Q_1(A;x,y)$. Then $P_1(A;x,y)$ or $P_1(A;y,x)$. Suppose $B,C \supseteq A$, $B \subseteq C$ and $P_0(B;x,y)$. Since $P_1(A;x,y)$ or $P_1(A;y,x)$ and P_1 is upward stable, $P_0(B;x,y)$ implies $P_1(B;x,y)$. Since P_1 is upward stable, $P_1(C;x,y)$ and thus $P_0(C;x,y)$. This shows that P_0 is upward stable for $\langle x,y \rangle$ from A when $\neg Q_1(A;x,y)$.

P_0 is downward stable for $\langle x,y \rangle$ from A for all A in $\mathbb{D}$ and all x,y in A such that $Q_1(A;x,y)$. To see this, suppose that x,y are in A and $Q_1(A;x,y)$. Suppose further that $B \subseteq C \subseteq A$, $x,y \in B$ and $P_0(C;x,y)$. Since Q_1 is downward stable, $Q_1(B;x,y)$ and $Q_1(C;x,y)$. $P_0(C;x,y)$ and $Q_1(C;x,y)$ implies $P_2(C;x,y)$. Since P_2 is stable, $P_2(B;x,y)$ and together with $Q_1(B;x,y)$ this implies $P_0(B;x,y)$. This shows that P_0 is downward stable for $\langle x,y \rangle$ from A when $Q_1(A;x,y)$.

We conclude this section by raising the following problems: Does P_0^* function as the raw data giver for P_0? The answer seems to be no. To see this, suppose that x,y and z are as above and compare them with x', y' and z' in U which are such that

$$P_1(\{x',y'\}; x',y') \;\&\; Q_1(\{y',z'\}; y',z') \;\&\; Q_1(\{x',z'\}; x',z'),$$
$$P_2(\{x',y'\}; y',x') \;\&\; P_2(\{y',z'\}; y',z') \;\&\; P_2(\{x',z'\}; z',x').$$

Hence,

$$P_0(\{x',y'\}; x',y') \;\&\; P_0(\{y',z'\}; y',z') \;\&\; P_0(\{x',z'\}; z',x').$$

This implies that

$$P_0^*(\{x',y',z'\}) = \{ \langle x',y' \rangle , \langle y',z' \rangle , \langle z',x' \rangle \} .$$

Since,

$$P_1(\{x',y',z'\}) = \{ \langle x',z' \rangle , \langle z',y' \rangle , \langle x',y' \rangle \}$$

it follows that

$$P_0(\{x',y',z'\}) = \{ \langle x',z' \rangle , \langle z',y' \rangle , \langle x',y' \rangle \}.$$

Note that
$$P_0^*(\{x,y,z\}) = \{ \langle x,y \rangle , \langle y,z \rangle , \langle z,x \rangle \} .$$
The analogy between $P_0^*(\{x,y,z\})$ and $P_0^*(\{x',y',z'\})$ can hardly be missed. The primed objects play the same role here as the corresponding unprimed object. To be more exact, the mapping
$$\varphi:\{x,y,z\} \rightarrow \{x',y',z'\}$$
such that
$$\varphi(x) = x', \;\; \varphi(y) = y', \;\; \varphi(z) = z'$$
is an isomorphism on $P_0(\{x,y,z\})$ to $P_0(\{x',y',z'\})$. However, φ is not an isomorphism on $P_0(\{x,y,z\})$ to $P_0(\{x',y',z'\})$ since for example $P_0(\{x,y,z\}; z,x)$ but $\neg P_0(\{x',y',z'\}; x',z')$. The primed objects do not therefore play the same role in this case as the corresponding unprimed objects. P_0 is thus not completely determined by P_0^* in the sense we will in the next chapter call subordination, i.e. P_0 is not subordinate to P_1^*. And it may seem natural to demand of a raw data giver that it determines the relational it is a raw data giver for. In the next chapter the notion of a relational being *subordinate* to its raw data giver will be developed.

3.4 Finitary systems of relationals

In set-theory, a finitary relational system or *structure* **A**, as we prefer to call it here, is a sequence of the form $\langle A, \rho_1,...,\rho_k \rangle$ where A, the base set (or domain) of the structure, is a non-empty set and $\rho_1,...,\rho_k$ are relations in A. We now introduce the corresponding notions for relationals. A *finitary system of relationals* **R** is a finite sequence $\langle R_1,...,R_k \rangle$ of relationals $R_1,...,R_k$ with the same range of definition $\mathbb{D}$. We call $R_1,...,R_k$ the constituents of the system. By analogy with our treatment of relationals, $\langle R_1,...,R_n \rangle$ will be regarded as a function. For all A in $\mathbb{D}$
$$\langle R_1,...,R_k \rangle(A) = \langle A, R_1(A),...,R_k(A) \rangle .$$
$\mathbb{D}$ is thus the range of definition of $\langle R_1,...,R_k \rangle$. The value of $\langle R_1,...,R_k \rangle$ for A is a relational structure with A as base set and containing the extensions of $R_1,...,R_k$ on A. As for relationals, we call the union of the elements in $\mathbb{D}$ the universe of application of the system.

We will often use the expression "relational system" rather than "system of relationals" for convenience.. It is therefore important that the reader remember the difference between relational structures and relational systems.

The notions of type originally applied to relational structures will here be applied even to relational systems. If R_i is n_i-ary then $\langle n_1,...,n_k \rangle$ is the *type* of the relational system $\langle R_1,...,R_k \rangle$.

A relational system will often be symbolized by **P,Q,R,S** or **T**, sometimes appropriately indexed. An n-ary relational R can be regarded as a relational system of type $\langle n \rangle$, i.e. we can identify R with $\langle R \rangle$. When we talk about an arbitrary relational system **R**, we shall normally not exclude the possibility that **R** is (or contains) just one relational R. The range of definition and the universe of application of a relational system **R** will often be denoted $\mathbb{D}_\mathbf{R}$ and $\mathbb{U}_\mathbf{R}$, respectively.

As the value of $\langle R_1,...,R_k \rangle$ for A, we take $\langle A,R_1(A),...,R_k(A) \rangle$ here and not $\langle R_1(A),...,R_k(A) \rangle$. The reason is obvious; if we include A we get a relational structure and otherwise just a k-tuple of relations. Let us call $\langle A,R_1(A),...,R_k(A) \rangle$ the *extension* of $\langle R_1,...,R_k \rangle$ on A. For relational systems we do not distinguish between extensions and graphs and prefer to use the word "extension". The extension of $\langle R \rangle$ is thus the graph of R. As was said before, the term "extension" will be used in later chapters for relationals too, it being clear from the context whether "the extension of R on A" is meant as a relation ρ or a relational structure $\langle A,\rho \rangle$.

Most of the concepts defined for relationals in the previous sections could easily be extended to relational systems. Let us first consider the different notions of stability. The system $\mathbf{R}=\langle R_1,...,R_k \rangle$ is stable if R_i is stable for all i, $1 \le i \le k$. We define weakly stable, upward stable and downward stable in an analogous way. Suppose that **R** is stable. Since

$$\mathbf{R}(A)/B = \langle A \cap B; R_1(A)/B,...,R_k(A)/B \rangle$$

it follows from the stability of R_i that

$$\mathbf{R}(A)/B = \langle A \cap B; R_1(B)/A,...,R_k(B)/A \rangle ,$$

i.e.

$$\mathbf{R}(A)/B = \mathbf{R}(B)/A .$$

Hence, if **R** is stable then for all A and B in $\mathbb{D}_\mathbf{R}$

$$\mathbf{R}(A)/B = \mathbf{R}(B)/A .$$

It is easy to see that the converse also holds. The definition of stability for relationals is therefore applicable directly even to relational systems. The situation is analogous for weak stability.

The extension class of the relational system **R** is the class

$$\{ \mathbf{R}(A) \mid A \in \mathbb{D}_\mathbf{R} \}$$

and we denote it $\mathbb{E}_\mathbf{R}$. If $\mathbf{R}=\langle R_1,...,R_k \rangle$ then

$$\mathbb{E}_{\langle R_1,...,R_k\rangle} = \{\langle A,R_1(A),...,R_k(A)\rangle \mid A \in \mathbb{D}_{\langle R_1,...,R_k\rangle}\}$$

Note that if $R = \langle R_1,...,R_k\rangle$ then

$$\langle A,\rho_1,...,\rho_k\rangle \in \mathbb{E}_R \text{ iff } \langle A,\rho_i\rangle \in \mathbb{E}_{R_i}$$

for all i, $1 \leq i \leq k$. The characteristic class of R is denoted $\mathbb{C}_R$ and

$$\mathbb{C}_R = \{X \mid \exists A \in \mathbb{D}_R : R(A) \cong X\} .$$

Note that

$$\mathbb{C}_{\langle R_1,...,R_k\rangle} \subseteq \{\langle X,\rho_1,...,\rho_k\rangle \mid \langle X,\rho_i\rangle \in \mathbb{C}_{R_i} \text{ for all } i, 1 \leq i \leq k\} .$$

The notions of extension theory and satisfaction of a set-theoretical predicate are straightforwardly extendable to relational systems.

A relational system is local if all its constituents are local. If $\langle n_1,...,n_k\rangle$ is the type of $\langle R_1,...,R_k\rangle$ and $1 \leq j \leq k$ and there is no n_i, $1 \leq i \leq k$, greater than n_j, then $\langle R_1,...,R_k\rangle$ is local iff R_j is local iff every set of n_j or fewer elements of U_{R_k} belongs to $\mathbb{D}_{\langle R_1,...,R_k\rangle}$. If $\langle R_1,...,R_k\rangle$ is local then the regionalization $\langle R_1,...,R_k\rangle^*$ exists and

$$\langle R_1,...,R_k\rangle^* = \langle R_1^*,...,R_k^*\rangle .$$

Hence,

$$\langle R_1,...,R_k\rangle^*(A) = \langle A,R_1^*(A),...,R_k^*(A)\rangle .$$

In chapter 7 we shall consider relational systems which are not necessarily finitary.

3.5 Historical and bibliographical remarks

The idea behind the notion of a relational is that it is meaningful to talk about the extension of a relation over a set or in a situation. This idea is not at all new. On the contrary, it seems to be an old idea which in different contexts has taken different forms. Let me give just a few examples where the idea has been used.

In *Meaning and necessity* and in lecture notes (Carnap, 1972), Rudolf Carnap suggested that certain intensional entities, in particular the intensions of linguistic expressions, be identified with functions that take possible state of affairs (or possible worlds) as arguments. The value of a function that serves as the intension of a linguistic expression at a world is the extension of the expression at that world. In his formal semantics, Carnap represented possible worlds by state descriptions and later by set theoretical structures (models).

Carnap's idea has been developed further by Kanger, Kripke, Kaplan and Montague among others. (See the first chapter of Gallin, 1975, for a survey of some of Carnap's and Montague's ideas.) In Montague (1969) it is argued that a property of individuals may be viewed as a function having the set I of all possible worlds as its domain and subsets of the set U of all possible individuals as values. Montague exemplifies this with the property of being red, which is identified with the function which assigns to each possible world the set of possible individuals which in that world are red. And Montague continues:

> More generally, an *n-place predicate of individuals* is a function having I as its domain and of which the values are sets of *n*-place sequences of members of U. If P is a predicate of individuals and i a possible world, we regard *P(i)* as the *extension* of P in i; the extension of an *n*-place predicate will always be an *n*-place relation (in the extensional sense). (Montague, 1974, p. 152.)

Relationals resemble Montague's predicates in that they are functions with extensions as values.

There are thus some formal similarities between relationals and some notions in theories of modal logic. The similarity with Kanger's version of the semantics for modal logic is perhaps most striking, since functions representing intensional objects there have domains as arguments. But there are also differences. The domains in the range of definition of a relational are not intended to represent possible worlds. Even in the actual world, a relational can have different extensions depending on what domain one considers. In chapter 4 the notion of a possible extension of a relational will be introduced, and a possible extension is a kind of a possible world —but usually a small one, since it characterizes a possible state of affairs with respect to the relational.

The idea of intensional entities as functions which have extensions as values was first used in formal semantics. The notion of a relational is intended to be used within measurement theory and related areas. It seems to me that by using relationals in measurement theory and in some areas of decision theory, one only says explicitly what is in a sense already said implicitly. But by introducing relationals explicitly it becomes possible to study certain phenomena (like stability etc.) that are not easily discussed within a framework where relations are simply represented as sets of ordered tuples.

The treatment of stability given both in this chapter and in the rest of the book is much influenced by the work done by Sven Danielsson and Margareta Sjöberg in the early 70's on Arrow's condition of independence of irrelevant alternatives in social choice theory. My interest in what are called relationals here arose out of discussions with them. Of special importance for the content of this chapter has been the application in Danielsson (1974) of the condition of independence or irrelevant alternatives to individual preferences and to other relations quite different from group preferences.

CHAPTER 4

SUBORDINATION, UNCORRELATION AND DERIVATION

4.0 Introduction

In this chapter we shall introduce three main concepts we are going to use in the study of structural dependence, viz. subordination, uncorrelation and derivation. Subordination is one end of the "dependence-scale"—complete determination—and uncorrelation the other end—complete undetermination. (To be exact, there are as we shall see in later chapters many "dependence-scales" and subordination and uncorrelation are the endpoints of one of them.) We start off in section 4.1 with a presentation of the core of subordination, that is isomorphism preservation. In that section we also introduce the notion of a transition, which will be one of our basic tools in the rest of the investigation. Section 4.2 is devoted to subordination and its relation to definability while uncorrelation is treated in section 4.3. In section 4.4 we show that R* is not generally subordinate to R. What is meant by equality, =, for relationals is discussed in section 4.5. There we also introduce the notion of a decision method for relationals. Decision methods are also the topic of section 4.6, now applied to derivations of one relational from another. That a relational is derivable from another turns out to be another kind of dependence than isomorphism preservation and its relatives. In section 4.7 the notions of stability, introduced in chapter 3 as properties of relationals, will be applied to transitions and derivations.

Both subordination and derivability seem to express some kind of dependence, but it is a difference; subordination is a kind of structural dependence in a sense which derivability is not. The structural character of transitions and subordination will be discussed in section 4.8

In section 4.9 a family of relevance notions defined in terms of subordination, correlation and transitions is introduced.

In chapter 7 we shall conduct a more detailed study of transitions, subordination and uncorrelation and this in a more general framework of systems of relationals eventually containing infinitely many relationals. The proofs of the elementary results of transitions, subordination and uncorrelation which are just stated in chapter 4 can be found in chapter 7.

In this and later chapters we disregard the difference between extension and graph and let R(A), which we call the extension of R on A, be either the relation ρ or the structure $\langle A, \rho \rangle$; it is almost always clear from the context what is intended. When we want to emphasize the distinction we call ρ the proper extension of R on A and $\langle A, \rho \rangle$ the graph of R on A. Note that R as the assignment of graphs to sets is a one-to-one function.

4.1 Isomorphism preservation and transitions

If the range of definition $\mathbb{D}$ of the relational R is a subset of the range of definition of the relational S, then we can construct the set
$$\mathbb{F}_{RS} = \{ \langle R(A), S(A) \rangle \mid A \in \mathbb{D} \} \ .$$
It is here essential that R(A) and S(A) are graphs. Note the similarity between $\mathbb{F}_{RS}$ and $\mathbb{E}_{\langle R, S \rangle}$; we have
$$\langle \langle X, \rho \rangle , \langle Y, \sigma \rangle \rangle \in \mathbb{F}_{RS} \text{ iff } \langle X, \rho, \sigma \rangle \in \mathbb{E}_{\langle R, S \rangle} \ .$$
$\mathbb{F}_{RS}$ is of course a function of $\mathbb{E}_R$ into $\mathbb{E}_S$ because if $R(A)=R(A')$ then, since R(A) is the graph of R on A, $A=A'$ and thus $S(A)=S(A')$. Hence, $S(A) = \mathbb{F}_{RS}(R(A))$ for all elements A in $\mathbb{D}$. An equivalent and often convenient way of expressing this is $S = \mathbb{F}_{RS} \circ R$. This means that the mere existence of a function f such that
$$S = f \circ R$$
does not imply any connection between R and S beyond the fact that $\mathbb{D}_R \subseteq \mathbb{D}_S$. The function f must have some additional properties to secure some essential connection between R and S. Such a property is isomorphism preservation, which we now introduce.

Let f be a function on a set K of similar structures of the type τ to a set K' of similar structures of the type τ'. Suppose further that the value of f for each argument **A** is a structure which has the same domain as **A**; f is thus domain-preserving. We call such a function an *s-function* ("s" for "structure"; it might alternatively be called a "transfiguration"). f is *isomorphism-preserving* if for every **A** and **B** in K, if φ is an isomorphism on **A** to **B**, then φ is also an isomorphism on f(**A**) to f(**B**). $\mathbb{F}_{RS}$ is thus isomorphism-preserving iff the following condition holds for all elements A and B in $\mathbb{D}$: If φ is an isomorphism on R(A) to R(B) then φ is an isomorphism on S(A) to S(B). (φ is an isomorphism on R(A) to R(B) iff φ is a one-to-one function on A onto B and for all $a_1,...,a_n$ in A

$$R(A; a_1,...,a_n) \text{ iff } R(B; \varphi(a_1),...,\varphi(a_n)).$$

(See section 2.9 for further details about isomorphisms.) The condition above does not explicitly mention $\mathbb{F}_{RS}$, so we apply the notion of isomorphism preservation directly to relationals. Let us say that S is *isomorphism-preserving* relative to R if $\mathbb{D}_R \subseteq \mathbb{D}_S$ and for all A and B in $\mathbb{D}_R$,

$$\mathbb{I}(R(A),R(B)) \subseteq \mathbb{I}(S(A),S(B)).$$

($\mathbb{I}(\mathbf{X},\mathbf{Y})$ is the set of isomorphisms of **X** to **Y**, see section 2.9.) It is easy to see that isomorphism preservation is a transitive and reflexive relation.

Isomorphism preservation is just one of a family of related concepts. Two other members of this family, automorphism preservation and homomorphism preservation are defined by: S is *automorphism-preserving* relative to R iff $\mathbb{D}_R \subseteq \mathbb{D}_S$ and for all A in $\mathbb{D}_R$,

$$\mathbb{I}(R(A)) \subseteq \mathbb{I}(S(A)).$$

S is *homomorphism-preserving* relative to R iff $\mathbb{D}_R \subseteq \mathbb{D}_S$ and for all A and B in $\mathbb{D}_R$,

$$\mathbb{H}o(R(A),R(B)) \subseteq \mathbb{H}o(S(A),S(B)).$$

where $\mathbb{I}(\mathbf{X})$ is the set of automorphisms of **X** and $\mathbb{H}o(\mathbf{X},\mathbf{Y})$ the set of homomorphisms of **X** onto **Y**.

Isomorphism preservation follows from homomorphism preservation but implies automorphism preservation and is therefore in a sense stronger than automorphism preservation, but it seems to be more fundamental. We shall now take a first look at the meaning of isomorphism preservation. To this end we introduce the notion of a transition between two relationals.

Suppose that the range of definition of R is a subset of the range of definition of S. Close $\{\langle R(A),S(A)\rangle \mid A \in \mathbb{D}_R\}$, i.e. $\mathbb{F}_{RS}$, under isomorphisms simultaneously for both components in the ordered pair, i.e. construct the class

$$\{\langle \mathbf{X},\mathbf{Y}\rangle \in \mathbb{C}_R \times \mathbb{C}_S \mid X=Y \ \& \ \exists A \in \mathbb{D}_R : \exists \varphi \in \mathbb{B}i(X,A) : \varphi \in \mathbb{I}(X,R(A)) \ \& \ \varphi \in \mathbb{I}(Y,S(A))\}.$$

($\mathbb{B}$i(X,A) is the set of bijections from X to A, see section 2.1.) Note that we take the same isomorphisms for both structures X and Y. This class is a subclass of $\mathbb{C}_R \times \mathbb{C}_S$ and is thus a correspondence on $\mathbb{C}_R$ to $\mathbb{C}_S$. (See section 2.6 for details on correspondences.) We call it *the transition from R to S* and denote it R^S. If the ranges of definition of R and S are the same, then R^S is a correspondence on $\mathbb{C}_R$ *onto* $\mathbb{C}_S$. Let us agree to regard R^S as defined only if $\mathbb{D}_R \subseteq \mathbb{D}_S$. Note that if R^S is the transition from R to S then $\langle R(A),S(A) \rangle \in R^S$, i.e. $R(A)R^S S(A)$, for all $A \in \mathbb{D}_R$. It is also easy to see that if R and S have the same range of definition, then

$$X R^S Y \text{ iff } Y S^R X.$$

To make the "meaning" of the transitions between relationals somewhat clearer the following informal considerations may be of help. Compare $\mathbb{F}_{RS} =$ $= \{\langle R(A),S(A) \rangle \mid A \in \mathbb{D}\}$ and R^S. $\mathbb{F}_{RS}$ "operates" on relational structures and for them the base set plays an important role. Through the construction of R^S we leave the base sets out of account and consider only the structure of the relational structures. It is important here to distinguish between the two uses of "structure". A relational structure is an ordered n-tuple whose first component is the base set of the structure. The structure or form of a relational structure can within a set-theoretical framework be regarded as the class of all relational structures isomorphic to the given one. And R^S is $\mathbb{F}_{RS}$ closed under isomorphisms, N.B. *simultaneously* for both the components. We shall develop this line of thought further and in a more fundamental way in section 4.8. For the moment let us just note that the correspondence R^S "operates" in a sense on the structure or form of relational structures. Transitions are therefore useful for studying structural dependence (the structural aspect of dependence).

What use can we make of the transition from R to S? The idea we shall investigate here is that R^S expresses or formulates the connection or correlation between R and S. The more extensive R^S is, the weaker the connection between R and S. R^S can at most be the class

$$\{\langle X,Y \rangle \in \mathbb{C}_R \times \mathbb{C}_S \mid X=Y\}$$

because if $X R^S Y$ then $X=Y$. On the other hand, when R^S is as restricted as possible, it is a one-to-one function on $\mathbb{C}_R$ into $\mathbb{C}_S$ (and onto $\mathbb{C}_S$ if $\mathbb{D}_R = \mathbb{D}_S$), because R^S has $\mathbb{C}_R$ as its domain. R^S is always situated between these two extremes. It is of special interest under which circumstances R^S is a function.

Correspondences are just binary relations, so one may question the point of emphasizing that transitions are correspondences rather than simply saying that they are binary relations. The reason is that the correspondence terminology stresses the "mapping character" of binary relations. A univalent correspondence is a mapping (function) and a bijective mapping is a one-to-one correspondence. (See section 2.6.) Since we are especially interested in the circumstances when the transition between two relationals are a function it is natural to describe the transitions as correspondences.

At this point, the following observation—which we formulate as a theorem—is important:

Theorem 4.1.1: The transition from R to S is a function iff S is isomorphism-preserving relative to R.

In the proof we shall use the following elementary properties of isomorphisms; see section 2.9 for details. What is meant by $\varphi[X]$ is explained in section 2.8.)

(i) $\varphi \in \mathbb{I}(X,Y)$ & $\psi \in \mathbb{I}(X,Z)$ implies $\psi \circ \varphi^{-1} \in \mathbb{I}(Y,Z)$.

(ii) $\varphi \in \mathbb{I}(X,Y)$ iff $\varphi \in \mathbb{B}i(X,Y)$ & $\varphi[X]=Y$

(iii) $\varphi \in \mathbb{I}(X,Y)$ & $\psi \in \mathbb{I}(Z,W)$ & $\psi \circ \varphi^{-1} \in \mathbb{I}(Y,W)$ implies $X=Z$.

We also use the following rather obvious fact, which is proved as proposition 7.2.4.

(iv) XR^SY and $\varphi \in \mathbb{B}i(X,Z)$ implies $\varphi[X]R^S\varphi[Y]$.

We divide the theorem into two parts and formulate each as a lemma.

Lemma 4.1.2: If S is isomorphism-preserving relative to R, then the transition from R to S is a function.

Proof: Let us suppose that S is isomorphism-preserving relative to R. Then $\mathbb{D}_R \subseteq \mathbb{D}_S$. Suppose further that XR^SY and XR^SZ. Then there are $A,B \in \mathbb{D}_R$ and $\varphi \in \mathbb{B}i(X,A)$ and $\psi \in \mathbb{B}i(X,B)$ such that

$$\varphi \in \mathbb{I}(X,R(A)) \ \& \ \varphi \in \mathbb{I}(Y,S(A))$$

and

$$\psi \in \mathbb{I}(X,R(B)) \ \text{and} \ \psi \in \mathbb{I}(Z,S(B)).$$

Note that $\psi \circ \varphi^{-1} \in \mathbb{B}i(A,B)$ and $\psi \circ \varphi^{-1} \in \mathbb{I}(R(A),R(B))$ according to (i). Because S is isomorphism-preserving relative to R, $\psi \circ \varphi^{-1} \in \mathbb{I}(S(A),S(B))$. Together with $\varphi \in \mathbb{I}(Y,S(A))$ and $\psi \in \mathbb{I}(Z,S(B))$ this implies, according to (iii), that $Y=Z$. Therefore R^S is univalent and thus a function. ♦

Lemma 4.1.3: If R^S is a function then R^S is isomorphism-preserving.

Proof: (I) Suppose that R^S is a function. Suppose further that AR^SA' and BR^SB' and $\varphi \in I(A,B)$. Hence, according to (iv) $\varphi[A]R^S\varphi[A']$ and according to (ii) $\varphi[A]=B$. We thus get $BR^S\varphi[A']$. Since R^S is a function and BR^SB', it follows that $\varphi[A'] = B'$. From (ii) then follows $\varphi \in I(A',B')$, i.e. $\varphi \in I(R^S(A),R^S(B))$. This proves that R^S is isomorphism-preserving. ◆

The theorem follows from lemma 4.1.2 and 4.1.3 as soon as we observe that if R^S is an isomorphism-preserving function then S is isomorphism-preserving relative to R. For if R^S is an isomorphism-preserving function, then $\varphi \in I(R(A),R(B))$ implies $\varphi \in I(S(A),S(B)$, since $R(A)R^SS(A)$ and $R(B)R^SS(B)$.

It is obvious that it holds generally that

$$\{\langle R(A),S(A)\rangle \mid A \in D\} \subseteq \{\langle X,Y\rangle \in R^S \mid X \in E_R\}.$$

The set inclusion can be sharpened to equality iff S is isomorphism-preserving relative to R. We formulate this observation as a lemma.

Lemma 4.1.4: $\{\langle R(A),S(A)\rangle \mid A \in D\} = \{\langle X,Y\rangle \in R^S \mid X \in E_R\}$
iff S is isomorphism-preserving relative to R.

Proof: (I) Suppose that S is isomorphism-preserving relative to R. It remains to prove that

$$\{\langle X,Y\rangle \in R^S \mid X \in E_R\} \subseteq F_{RS}.$$

Suppose $\langle X,Y\rangle \in R^S$ and $X \in E_R$. Then $X=R(X)$ and there are $A \in D$ and $\varphi \in Bi(X,A)$ such that $\varphi \in I(X,R(A))$ and $\varphi \in I(Y,S(A))$. Hence, $\varphi \in I(R(X),R(A))$. Since S is isomorphism-preserving relative to R it holds that $\varphi \in I(S(X),S(A))$. We have therefore $\varphi[Y]=S(A)$ and $\varphi[S(X)]=S(A)$, and thus $\varphi[Y]=\varphi[S(X)]$. From this follows that $Y=S(X)$ and hence $\langle X,Y\rangle=\langle R(X),S(X)\rangle$, and $\langle R(X),S(X)\rangle$ obviously belongs to F_{RS}.

(II) Suppose that

$$F_{RS} = \{\langle X,Y\rangle \in R^S \mid X \in E_R\}.$$

Suppose further that $\varphi \in I(R(A),R(B))$. From $R(A)R^SS(A)$ and φ a bijection follows according to (iv)

$$\varphi[R(A)]R^S\varphi[S(A)]$$

(This is proved in section 7.2). $\varphi[R(A)]=R(B)$, so $\varphi[R(A)]$ belongs to E_R. According to the assumption it then follows that $\langle \varphi[R(A)],\varphi[S(A)] \rangle \in F_{RS}$. Since

$\mathbb{F}_{RS}$ is a function and $\varphi[R(A)]=R(B)$ and $\langle R(B),S(B)\rangle \in \mathbb{F}_{RS}$, it follows that $\varphi[S(A)]=S(B)$. Hence, $\varphi \in \mathbb{I}(S(A),S(B))$. ◆

From theorem 4.1.1 and lemma 4.1.4 it follows that if S is isomorphism-preserving relative to R then R^S is a function on $\mathbb{C}_R$ into $\mathbb{C}_S$ and an extension of the function $\mathbb{F}_{RS}$ on $\mathbb{E}_R$ into $\mathbb{E}_S$. The extension of $\mathbb{F}_{RS}$ to R^S is very natural. To see this suppose that X is an element of $\mathbb{C}_R$. Then there is an A in $\mathbb{D}_R$ and a bijection φ on A onto X such that $X=\varphi[R(A)]$. Hence,
$$R^S(X)=R^S(\varphi[R(A)])=\varphi[S(A)]=\varphi[\mathbb{F}_{RS}(R(A))]=\varphi[\mathbb{F}_{RS}(\varphi^{-1}[X])].$$
The figure below illustrates the situation.

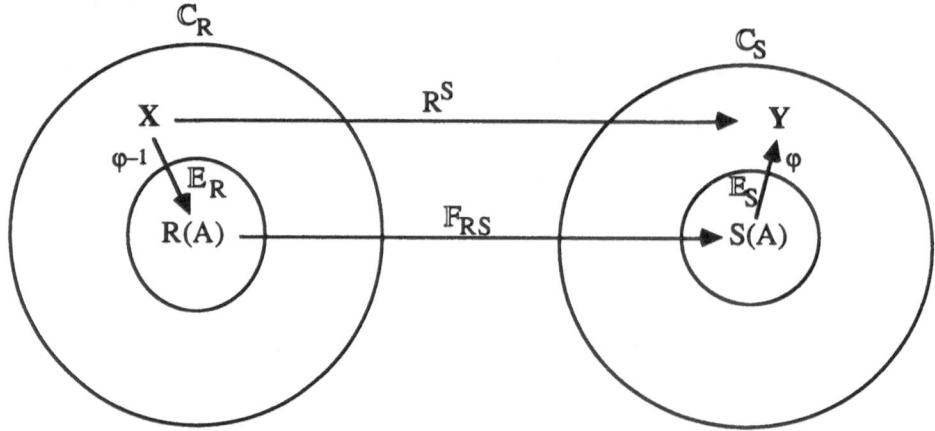

$R^S(X) = \varphi[\mathbb{F}_{RS}(\varphi^{-1}[X])]$ where φ an isomorphism from R(A) to **X**.

Note that it holds generally, i.e. even when R^S is not certainly a function, that $XR^S\varphi[\mathbb{F}_{RS}(\varphi^{-1}[X])]$ where $\varphi \in \mathbb{B}i(X,R(A))$.

That S is isomorphism-preserving relative to R implies a strong connection between S and R. It is, as we have seen, equivalent to the following condition: For all A in $\mathbb{D}_R$ S(A) is the only structure such that $R(A)R^SS(A)$. Hence, if we presuppose that $\mathbb{D}_R \subseteq \mathbb{D}_S$, then it holds that S is not isomorphism-preserving relative to R iff there is A in $\mathbb{D}_R$ such that
$$R(A)R^SS(A) \quad \& \quad R(A)R^SA \quad \& \quad A\neq S(A).$$
We can therefore say that if S is isomorphism-preserving relative to R, then S(A) is determined by R(A) through R^S. As we have seen R^S operates in a sense on the structure or form of the extensions of R. S(A) can therefore be said to be deter-

mined by R(A) for structural reasons; it is the structure or form on R(A) that determines S(A). The connection between S and R when S is isomorphism-preserving is therefore a kind of "structural determination". We shall return to this in section 4.8.

4.2 Subordination and definability

Isomorphism preservation is defined in such a way that if S is isomorphism-preserving relative to R then the range of definition of R is a subset of the range of definition of S. If S is isomorphism-preserving relative to R and the range of definition of S is the same as the range of definition of R, then we say that S is *subordinate* to R. If S is subordinate to R then, in an intuitive sense, all the information contained in S is already contained in R and R has in a sense at least as much expressive power as S. Reciprocal subordination will be called *parity*, i.e. if S is subordinate to R and R is subordinate to S, then S and R are *on a par*. If R and S are on a par, then R and S contain—in an intuitive sense—the same information and can be said to have the same expressive power. Note that subordination, like isomorphism preservation, is a transitive and reflexive relation and that parity is an equivalence relation.

As the reader has presumably already realized, subordination is closely related to definability. Let us look at this relationship in more detail. First some preliminaries.

Suppose R and S have the same range of definition $\mathbb{D}$. Let us for convenience denote the extension class and the characteristic class of $\langle R,S \rangle$ by $\mathbb{E}_{RS}$ and $\mathbb{C}_{RS}$, respectively. Then

$$\mathbb{E}_{RS} = \{ \langle A,R(A),S(A) \rangle \mid A \in \mathbb{D} \}$$
$$\mathbb{C}_{RS} = \{ \langle X,\rho,\sigma \rangle \mid \exists A \in \mathbb{D}: \exists \varphi \in \mathbb{B}i(X,A): \varphi \in \mathbb{I}(\langle X,\rho,\sigma \rangle, \langle A,R(A),S(A) \rangle) .$$

Note that

$$\{ \langle X,\rho \rangle \mid \exists \sigma: \langle X,\rho,\sigma \rangle \in \mathbb{C}_{RS} \} = \mathbb{C}_R$$
$$\{ \langle X,\sigma \rangle \mid \exists \rho: \langle X,\rho,\sigma \rangle \in \mathbb{C}_{RS} \} = \mathbb{C}_S$$
$$\mathbb{C}_{RS} \subseteq \{ \langle X,\rho,\sigma \rangle \mid \langle X,\rho \rangle \in \mathbb{C}_R \ \& \ \langle X,\sigma \rangle \in \mathbb{C}_S \}$$
$$\langle X,\rho,\sigma \rangle \in \mathbb{C}_{RS} \text{ iff } \langle \langle X,\rho \rangle, \langle X,\sigma \rangle \rangle \in R^S .$$

From the last proposition and the fact that R^S is isomorphism-preserving when S is isomorphism-preserving relative to R (according to (4.1.2) and (4.1.3)), it follows that if S is subordinate to R then the following condition is satisfied:
For every $\langle X,\rho,\sigma \rangle$ and $\langle X',\rho',\sigma' \rangle$ in $\mathbb{C}_{RS}$, if φ is an isomorphism on $\langle X,\rho \rangle$ onto

$\langle X',\rho'\rangle$ then φ is an isomorphism on $\langle X,\sigma\rangle$ onto $\langle X',\sigma'\rangle$.

The converse holds, too, since it follows from the above condition that for all A,B in $\mathbb{D}$

$$\text{if } \varphi\in\mathbb{I}(R(A),R(B)) \text{ then } \varphi\in\mathbb{I}(S(A),S(B))$$

(because $\langle A,R(A),S(A)\rangle$ is an element in $\mathbb{C}_{RS}$). So S is isomorphism-preserving relative to R.

Let us now take a closer look at the relationship between subordination and definability. Suppose that $\mathbb{T}_{RS}$ is a theory that has $\mathbb{C}_{RS}$ as its class of models, with R and S as primitive symbols where the interpretation of R is the first (and that of S the second) relation in a model of $\mathbb{T}_{RS}$. If α is an interpretation of $\mathbb{T}_{RS}$ then R^α will, as usual, be used for the interpretation of R and S^α for the interpretation of S. $|\alpha|$ denotes the domain of α.

As is easily seen from the discussion above, S is subordinate to R if and only if the following holds:

(*)   If α and β are models of $\mathbb{T}_{RS}$ such that φ is an isomorphism on R^α to R^β, then φ is an isomorphism on S^α to S^β (i.e. an isomorphism on α to β).

It is a well-known fact that (*) equivalent to

(**)   If α and β are models of $\mathbb{T}_{RS}$ such that $|\alpha|=|\beta|$ and $R^\alpha=R^\beta$ then $S^\alpha=S^\beta$.

According to Padoa's principle (or method) (see Suppes, 1957, p. 169f or Stoll, 1963, p. 244), S is not definable in terms of R in $\mathbb{T}_{RS}$ if (**) is not the case. Thus if S is definable in terms of R in $\mathbb{T}_{RS}$ then S is subordinate to R. This implies that the subordination of S to R is a necessary condition for S being definable in terms of R. If $\mathbb{T}_{RS}$ is a first-order theory, then it is also a sufficient condition, according to Beth's theorem on definability.

If a primitive symbol P is definable in terms of the other primitive symbols in a theory, then P is often called dependent (as a symbol of the theory). But if P is definable in terms of other primitive symbols, P is completely determined by those, so it is in a certain sense more adequate to say that P is completely (or wholly) dependent. It is convenient for the purpose of this study to give the concept of complete dependence an algebraic definition and not connect it with definability which presupposes considerations of the language used. Let $\mathbb{T}$ be a theory with two primitive symbols P and Q. We say that Q is *completely dependent* on P in $\mathbb{T}$ iff

(∗∗∗) if α and β are models of T such that φ is an isomorphism on P^α to P^β, then φ is an isomorphism on Q^α to Q^β (i.e. φ is an isomorphism on α to β).

S is thus subordinate to R iff S is completely dependent on $\mathbf{R}$ in T_{RS}, where T_{RS} is a theory which has $\mathbb{C}_{RS}$ as its class of models. The relational S could therefore be called completely dependent on the relational R if S is subordinate to R. But we shall not adopt this usage here, using instead "subordination" for the technical, well-defined concept applying to relationals and reserving "dependence" in connection with relationals (supplemented with appropriate attributes) for the informal discussion of the motivation and intended application of the technical concepts.

In the next section we will look at some examples which illustrate the notion of subordination.

4.3   Uncorrelation

Subordination is a strong form of dependence between two relationals. S is subordinate to R iff for all $\mathbf{A}$ in $\mathbb{C}_R$ there exists only one $\mathbf{B}$ in $\mathbb{C}_S$ such that $A R^S B$. Thus, given the extension of R on A then there is only one possible extension of S on A. We can therefore say that S is subordinate to R iff S is (in a certain sense) completely dependent on and wholly determined by R. If a relation S is not completely dependent on R it might still be dependent on R in a weaker sense. Consequently, there are weaker forms of dependence between two relationals than subordination. A good start for the study of weaker forms of dependence may be to ask what "complete independence" should mean applied to relationals. What does it mean that S is completely independent of and wholly undetermined by R? A look at the transition from R to S will give the answer: S is completely independent of R iff R^S is as extensive as possible. The technical concept introduced to catch this idea will be called "uncorrelation" and the exact definition is as follows: S is *uncorrelated* with R iff

$$R^S = \{\langle X, Y \rangle \in \mathbb{C}_R \times \mathbb{C}_S \mid X = Y\} .$$

We can express the definition in terms of characteristic classes instead of correspondences: S is *uncorrelated* with R iff

$$\mathbb{C}_{RS} = \{\langle A, \rho, \sigma \rangle \mid \langle A, \rho \rangle \in \mathbb{C}_R \text{ and } \langle A, \sigma \rangle \in \mathbb{C}_S\} .$$

It is now obvious what "correlation" means: S is *correlated* with R iff

$$R^S \subset \{\langle X, Y \rangle \in \mathbb{C}_R \times \mathbb{C}_S \mid X = Y\}$$

or, equivalently

$$C_{RS} \subset \{\langle A,\rho,\sigma\rangle \mid \langle A,\rho\rangle \in C_R \text{ and } \langle A,\sigma\rangle \in C_S\} \ .$$

If uncorrelation is a formulation of the idea of complete independence, then correlation seems to be a formulation of the idea of incomplete independence, which is the weakest form of dependence (within the actual context). If we take "dependence" to mean dependence to some degree (no matter how small), then correlation seem to be the right explication.

It might be appropriate here to look at a few examples.

(I) Let R be a weak ordering with the range of definition $\mathbb{D}$ and the universe of application $\mathbb{U}$ and $\mathbb{D}$ consisting of all non-empty subsets of $\mathbb{U}$. We further suppose that every weak order $\langle X,\rho\rangle$ is an element of C_R. Since R is a weak ordering it is true that for all A in $\mathbb{D}$ and all x,y,z in A the following two propositions hold:

$$xR(A)y \text{ and } yR(A)z \text{ implies } xR(A)z$$
$$xR(A)y \text{ or } yR(A)x \ .$$

We now define the equivalence relation S and the strict weak ordering P corresponding to R in the following way (see lemma 2.4.1): For all A in $\mathbb{D}$

(i) $xS(A)y$ iff $xR(A)y \ \& \ yR(A)x$

(ii) $xP(A)y$ iff $xR(A)y \ \& \ \neg yR(A)x$.

S is obviously subordinate to P, because S is definable in terms of P:

(iii) $xS(A)y$ iff $\neg xP(A)y \ \& \ \neg yP(A)x$.

To every extension of P there corresponds, therefore, exactly one extension of S. P is, however, not subordinate to S, since there is not a unique graph of P corresponding to every graph of S. Suppose for example that $A=\{a,b,c\}$ and that $S(A)$ is σ_1, where

$$\sigma_1 = \{ \ \langle a,a\rangle \ , \ \langle b,b\rangle \ , \ \langle c,c\rangle \ , \ \langle a,b\rangle \ , \ \langle b,a\rangle \ \} \ .$$

Then $P(A)$ has two possible values, viz. π_1 and π_2, where

$$\pi_1 = \{ \ \langle a,c\rangle \ , \ \langle b,c\rangle \ \}$$
$$\pi_2 = \{ \ \langle c,a\rangle \ , \ \langle c,b\rangle \ \}.$$

We have thus $\langle A,\sigma_1\rangle SP\langle B,\pi_1\rangle$ and $\langle A,\sigma_1\rangle SP\langle B,\pi_2\rangle$ where SP is the transition from S to P, and SP is therefore not a function. Hence, P is not subordinate to S. But P is correlated with S. Suppose $S(A)=\sigma_1$ as above. This implies a limitation on what the extension of P on A could possibly be. π_3, where

$$\pi_3 = \{ \ \langle a,b\rangle \ , \ \langle b,c\rangle \ , \ \langle a,c\rangle \ \}$$

is for example not a possible extension of P on A given that $S(A)=\sigma_1$, because $aS(A)b$ and this contradicts $aP(A)b$. Although $\langle A,\pi_3\rangle$ is an element of C_P,

$\langle A,\sigma_1,\pi_3 \rangle$ is therefore not an element of $\mathbb{C}_{SP}$ (the characteristic class of S and P). This implies that $\mathbb{C}_{SP}$ is not equal to

$$\{\langle A,\sigma,\pi \rangle \mid \langle A,\sigma \rangle \in \mathbb{C}_S \text{ and } \langle A,\pi \rangle \in \mathbb{C}_P\}$$

and that P is not uncorrelated with S. Even though P is not determined by S it is partially determined by—and to some degree dependent on—S.

(II) Let us now suppose that R is a preordering, i.e. transitive and reflexive, such that every preorder $\langle X,\rho \rangle$ is an element of $\mathbb{C}_R$. Define S and P as before by (i) and (ii). Then neither is P subordinate to S nor S to P. To see that S is not subordinate to P, let A,σ_1,π_1 and π_3 be as above and suppose $R(A)=\langle A,\rho_1 \rangle$ where

$$\rho_1 = \{ \langle a,c \rangle , \langle b,c \rangle , \langle a,b \rangle , \langle b,a \rangle , \langle a,a \rangle , \langle b,b \rangle , \langle c,c \rangle \} .$$

Then $S(A)=\langle A,\sigma_1 \rangle$ and $P(A)=\langle A,\pi_1 \rangle$. Suppose now that, rather than being equal to $\langle A,\rho_1 \rangle$, R(A) equals $\langle A,\rho_2 \rangle$ where

$$\rho_2 = \{ \langle a,c \rangle , \langle b,c \rangle , \langle a,a \rangle , \langle b,b \rangle , \langle c,c \rangle \} .$$

Then P(A) still equals $\langle A,\pi_1 \rangle$ but S(A) equals $\langle A,\sigma_2 \rangle$ where

$$\sigma_2 = \{ \langle a,a \rangle , \langle b,b \rangle , \langle c,c \rangle \} .$$

We thus have both $\langle A,\pi_1 \rangle P^S \langle A,\sigma_1 \rangle$ and $\langle A,\pi_1 \rangle P^S \langle A,\sigma_2 \rangle$, so P^S is not a function. If we had defined S by (iii) above, then S would of course have been subordinate to P. Note that P is correlated with S and S with P since it is not possible that $S(A)=\langle A,\sigma_1 \rangle$ and at the same time $P(A)=\langle A,\pi_3 \rangle$ and therefore $\langle A,\sigma_1,\pi_3 \rangle$ does not belong to $\mathbb{C}_{SP}$ and $\langle A,\pi_3,\sigma_1 \rangle$ does not belong to $\mathbb{C}_{PS}$.

(III) Let R_1 and R_2 be weak orders with the same range of definition $\mathbb{D}$, and such that the equality relation S_1, defined from R_1 according to (i) as before, is the same as the equality relation S_2 defined from R_2 according to (i). It is then possible that neither of P_1 and P_2 defined, respectively, from R_1 and R_2 according to (ii) above is subordinate to the other. But P_1 and P_2 are necessarily correlated. To see this, note that if $xP_1(A)y$, then it is not possible that $\neg xP_2(A)y$ and simultaneously $\neg yP_2(A)x$. For this implies $xS_2(A)y$ according to (iii) which is equivalent to $xS_1(A)y$ and contradicts $xP_1(A)y$.

(IV) The fourth and last example is as follows. Suppose that R and S have the same range of definition $\mathbb{D}$, which is the family of all non-empty subsets of a set $\mathbb{U}$, and that the theory $\mathbb{T}_{RS}$, which has $\mathbb{C}_{RS}$ as its set of models, has the following axioms:

1. xRx for all x
2. If xRy and yRx, then xRz
3. If xSy and ySz, then xSz

4. Exactly one of the following: xSy, ySx and xRy .

S is not subordinate to R. To see this, note that the condition (∗) in section 4.2 is not fulfilled by $\mathbb{T}_{RS}$. The structures $\alpha = \langle A, \rho_1, \sigma_1 \rangle$ and $\beta = \langle A, \rho_1, \sigma_2 \rangle$ where

$$A = \{1,2\} \qquad \rho_1 = \{\langle 1,1 \rangle, \langle 2,2 \rangle\} \qquad \sigma_1 = \{\langle 1,2 \rangle\} \qquad \sigma_2 = \{\langle 2,1 \rangle\}$$

are models of $\mathbb{T}_{RS}$ and the identity mapping ι on $\{1,2\}$ is an isomorphism from $\mathbf{R}^\alpha$ to $\mathbf{R}^\beta$ (i.e. from ρ_1 to ρ_1) but of course not an isomorphism from $\mathbf{S}^\alpha$ to $\mathbf{S}^\beta$ (i.e. from α to β). This is essentially an example in Suppes (1957) (see p. 169f.).

Is S correlated with R? It is rather obvious that the answer is yes, because axiom 4 connects the relationals R and S. A more formal argument runs as follows: Let γ be the structure $\langle A, \rho_2, \varnothing \rangle$ where A is as above, $\varnothing$ is the empty set and

$$\rho_2 = \{ \langle 1,1 \rangle , \langle 2,2 \rangle , \langle 1,2 \rangle , \langle 2,1 \rangle \}$$

Then γ is a model of $\mathbb{T}_{RS}$. Consequently, $\langle A, \varnothing \rangle$ is an element in $\mathbb{C}_S$. From the fact that α is a model of $\mathbb{T}_{RS}$ it follows that $\langle A, \rho_1 \rangle$ is an element in $\mathbb{C}_R$. But $\langle A, \rho_1, \varnothing \rangle$ is not a model of $\mathbb{T}_{RS}$ and therefore not an element in $\mathbb{C}_{RS}$. S is thus correlated with R.

Subordination has an analogue for primitive symbols in the actual extension theory and that is the condition (∗∗∗) in section 4.2. Also correlation has an analogue for primitive symbols. Let $\mathbb{T}$ be a theory with two primitive symbols $\mathbf{P}$ and $\mathbf{Q}$. We say that $\mathbf{Q}$ is *completely independent* of $\mathbf{P}$ in $\mathbb{T}$ iff for every model α of $\mathbb{T}$, if β is a model of $\mathbb{T}$ with the same domain as α then there is a model γ of $\mathbb{T}$ with the same domain as α such that $\mathbf{P}^\gamma = \mathbf{P}^\alpha$ and $\mathbf{Q}^\gamma = \mathbf{Q}^\beta$. Equivalently, $\mathbf{Q}$ is completely independent of $\mathbf{P}$ iff for every model α of $\mathbb{T}$,

$$\{\mathbf{Q}^\beta \mid \beta \text{ is a model of } \mathbb{T}, |\beta| = |\alpha| \text{ and } \mathbf{P}^\beta = \mathbf{P}^\alpha\} =$$
$$= \{\mathbf{Q}^\beta \mid \beta \text{ is a model of } \mathbb{T} \text{ and } |\beta| = |\alpha| \} .$$

Consequently, $\mathbf{Q}$ is not completely independent of $\mathbf{P}$ in $\mathbb{T}$ iff

$$\{\mathbf{Q}^\beta \mid \beta \text{ is a model of } \mathbb{T}, |\beta| = |\alpha| \text{ and } \mathbf{P}^\beta = \mathbf{P}^\alpha\} \subset$$
$$\subset \{\mathbf{Q}^\beta \mid \beta \text{ is a model of } \mathbb{T} \text{ and } |\beta| = |\alpha| \}.$$

The expected result that S is correlated with R iff S is not completely independent of $\mathbf{R}$ in $\mathbb{T}_{RS}$ holds, which easily can be shown.

Subordination is, as was pointed out in section 4.1 a reflexive and transitive relation, i.e. a preordering. I shall now mention some properties of correlation which are easy to prove; the proofs are omitted.

Both correlation and uncorrelation are symmetric, but neither is transitive. It is also the case that they are neither reflexive nor irreflexive. But under a general condition correlation is reflexive while uncorrelation is irreflexive, which is what would be expected. This is studied in some detail in section 7.3.

It may seem obvious that subordination implies correlation. But things are a little more complicated than that. The n-ary universal relational V defined by $V(A)=A^n$ for all A in $\mathbb{D}$ and the n-ary empty relational $\emptyset$ defined by $\emptyset(A)=\emptyset$ for all A in $\mathbb{D}$ are subordinate to as well as uncorrelated with every relational which has $\mathbb{D}$ as its range of definition. However, if there are $\underline{A}$ and $\underline{A}'$ in $\mathbb{C}_S$ such that A=A' but $\underline{A}\neq\underline{A}'$, then S is correlated with R if S is subordinate to R. The prerequisite that there exist $\underline{A}$ and $\underline{A}'$ in $\mathbb{C}_S$ such that $\underline{A}\neq\underline{A}'$ but A=A' is of course very general. If it holds we say that $\mathbb{C}_S$ is non-degenerate. Note that because of the symmetry of correlation R is correlated with S if S is subordinate to R, given that $\mathbb{C}_S$ is non-degenerate. (For further details see section 7.3.)

There are two equivalent characterizations of subordination which are both natural but of rather different kinds. One is by isomorphism preservation, i.e. in terms of classes of isomorphisms, and is in fact a kind of invariance condition. The other is by properties of the transition, more exactly, that it is univalent. The definition of correlation at the beginning of this section is by means of properties of the transition. It is therefore natural to ask if the condition on correlation could be formulated in terms of the isomorphism concept, i.e. if it is an invariance condition.

This question is related to the following: To what extent can correlation be analysed by means of subordination? The examples of correlation and subordination in this section illustrates roughly what is intended. If S is correlated but not subordinate to R it can for example be an effect of (1) R being subordinate to S (see example (I) above) or (2) the existence of a relational Q subordinate to S which is also subordinate to R (see example (III)). What other possibilities are there? We shall return to this question.

4.4 The dependence between R and its regionalization R*

In section 3.2 we introduced the regionalization-operation * for relationals. If the n-ary relational R is local then
$$R^*(A) = \{\langle x_1,...,x_n\rangle \mid x_1,...,x_n\in A \ \& \ R(\{x_1,...,x_n\};x_1,...,x_n)\}$$
for all A in $\mathbb{D}$. R* is thus a relational with the same range of definition as R. R* is of course in some sense defined or constructed from R. But note that R* is not

necessarily subordinate to R. The following simple example shows this.

Suppose that R is local and a strict weak ordering and a regional semiordering. Suppose further that R is the strict weak ordering associated with R*. Let $A=\{a_1,a_2,a_3\}$, $B=\{b_1,b_2,b_3\}$, where A and B be in $\mathbb{D}$ and further

$$R*(A) = \{ \langle a_1,a_2\rangle , \langle a_1,a_3\rangle , \langle a_2,a_3\rangle \}$$
$$R*(B) = \{ \langle b_1,b_3\rangle \}.$$

Then it holds that

$$R(A) = \{ \langle a_1,a_2\rangle , \langle a_1,a_3\rangle , \langle a_2,a_3\rangle \}$$
$$R(B) = \{ \langle b_1,b_2\rangle , \langle b_1,b_3\rangle , \langle b_2,b_3\rangle \} .$$

$\varphi:A\to B$ defined by $\varphi(a_i)=b_i$, $1\le i\le 3$, is an isomorphism from R(A) to R(B) but not an isomorphism from R*(A) to R*(B). So $\mathbb{I}(R(A),R(B))$ is not a subset of $\mathbb{I}(R*(A),R*(B))$ and thus R* is not subordinate to R. However, in this special case R is subordinate to R*. For if X and Y in $\mathbb{D}$ such that φ is an isomorphism on R*(X) to R*(Y) then φ is an isomorphism on R(X) to R(Y), since R(X) and R(Y) is the strict weak order associated with R*(X) and R*(Y), In the general case R is obviously not subordinate to R*.

The above example shows that for an arbitrary relational R, R* is not necessarily subordinate to R. But intuitively there seems, even in the general case, to be some kind of connection between R* and R. It is therefore natural to ask if R* is correlated with R. Under a general condition the answer is yes. Let us take a closer look at this. First a preliminary remark.

Let R be an n-ary local relational. Suppose X is a set consisting of n distinct elements $x_1,...,x_n$. Suppose further that $X_1= \langle X,\rho\rangle$ is an element in $\mathbb{C}_R$ and $X_2==\langle X,\sigma\rangle$ is an element in $\mathbb{C}_{R*}$ such that $\langle X,\rho\rangle R R^*\langle X,\sigma\rangle$. It then holds that

$$\rho(x_1,...,x_n) \text{ iff } \sigma(x_1,...,x_n) ,$$

which can be seen in the following way. There are $A\in \mathbb{D}$ and $\varphi\in \mathbb{B}i(X,A)$ such that $\varphi\in\mathbb{I}(X_1,R(A))$ and $\varphi\in\mathbb{I}(X_2,R*(A))$. Denote for each i, $1\le i\le n$, $\varphi(x_i)$ by a_i. Hence, $A=\{a_1,...,a_n\}$ where $a_1,...,a_n$ are distinct elements. According to the definition of R* we have

$$R*(A) = \{\langle y_1,...,y_n\rangle \,|\, y_1...y_n\in A \;\&\; R(\{y_1,...,y_n\}; y_1,...,y_n)\} .$$

From this follows

$$R(A; a_1,...,a_n) \text{ iff } R*(A; a_1,...,a_n) .$$

Since $\varphi^{-1}\in\mathbb{I}(R(A),X_1)$ and $\varphi^{-1}\in\mathbb{I}(R*(A),X_2)$ we get

$$\rho(\varphi^{-1}(a_1),...,\varphi^{-1}(a_n)) \text{ iff } \sigma(\varphi^{-1}(a_1),...,\varphi^{-1}(a_n)),$$

i.e.

$$\rho(x_1,...,x_n) \text{ iff } \sigma(x_1,...,x_n).$$

It is now easy to see that if there are $\langle X,\sigma \rangle$, $\langle X,\sigma' \rangle \in \mathbb{C}_{R*}$ such that X consists of n distinct elements $x_1,...,x_n$ and $\sigma(x_1,...,x_n)$ but $\neg\sigma'(x_1,...,x_n)$, then R* is correlated with R. For there is no $\langle X,\rho \rangle \in \mathbb{C}_R$ such that $\langle X,\rho \rangle R^{R*} \langle X,\sigma \rangle$ and $\langle X,\rho \rangle R^{R*} \langle X,\sigma' \rangle$; if there were such an element in $\mathbb{C}_R$ we should have, according to the remark above,

$$\rho(x_1,...,x_n) \text{ iff } \sigma(x_1,...,x_n)$$

and

$$\rho(x_1,...,x_n) \text{ iff } \sigma'(x_1,...,x_n),$$

which would contradict the supposition that $\sigma(x_1,...,x_n)$ but $\neg\sigma'(x_1,...,x_n)$. Hence,

$$R^{R*} \subset \{\langle Y,Y' \rangle \in \mathbb{C}_R \times \mathbb{C}_{R*} \mid Y=Y'\}$$

which shows that R is correlated with R*.

It may seem strange that R* is not subordinate to R for an arbitrary relational R. As has been emphasized before, subordination can be viewed as an explication of complete dependence—dependence in the sense of determination—between relationals. R* is obviously in the same sense constructed or defined from R since according to the definition of R*

$$R*(A) = \{\langle x_1,...,x_n \rangle \mid x_1,...,x_n \in A \ \& \ R(\{x_1,...,x_n\}; x_1,...,x_n)\}.$$

And so one would expect R* to be determined by R. In a sense it is. But it is not subordinate to R. It is at this point appropriate to emphasize an important property of subordination: it expresses *domainwise* dependence. If S is subordinate to R then S(A) is determined by the structure (or form) of R(A). In the example above R*(A) is not determined by the structure of R(A), so R* is not completely domainwise dependent on R. But it is still possible that R*(A) is determined by the structure of some other set of R-data than the extension of R on A. In fact, that is the case, so R* is completely (structurally) determined by R but not in a domainwise fashion.

We shall in this essay be mainly interested in domainwise dependence, but in chapter 9 we shall also begin the study of dependence which is not domainwise. In that chapter we shall return to the kind of dependence between R and R* which holds in general.

As has been pointed out earlier in this section R is not necessarily subordinate to R*. But the subordination of R to R* seems to be a necessary condition for R* providing the raw data for R. We have already touched upon this topic in the last example in section 3.3. The idea is that if R* is providing the raw data for R, then R(A) is completely determined by the structure of R*(A) for all A in $\mathbb{D}$, so R is structurally determined by R*.

4.5 Equality and decision methods for relationals

A relational R is of course a function; it takes sets as arguments and structures as values with the intended interpretation that R(A) is the extension of R on A. However, I have not taken the position here that a relational is necessarily a function in the set-theoretical sense. I prefer to think of relationals as a kind of "rule" or "programme". The reader may think of them as he likes, but must note how equality, = , for relationals is used here.

Two relationals R and S are said to be equal iff $\mathbb{D}_R = \mathbb{D}_S$ and for every A in $\mathbb{D}_R$, R(A)=S(A). This is in accordance with how equality is defined for set-theoretical functions. But I do not want to say that R and S are necessarily the same even if they are equal; equality does not imply identity. This use of equality is not new. Frege, for example, regarded the functions $f(x)=x^2-4x$ and $g(x)=x(x-4)$ as different but with equal value-ranges (Wertverlauf). (See Frege, 1952, p. 26.)

Most properties of relationals which we study in this essay are extensional in the sense that if R has a property and R=S (i.e. R is equal to S) then S has the property, too. But there is one notable exception and that is the concept of derived relation, which will be introduced in the next section.

In mathematical logic a function F is said to be effective if there is a decision method for F. A decision method for a function F is a method by which we can, given an argument a, calculate F(a) in a finite number of steps. Since a relational R is a function we may ask if R is an effective function. A decision method for R would therefore literally be a method such that given a domain A for R we can calculate R(A) in a finite number of steps. But if A is infinite then the graph R(A) is a structure with an infinite set as domain and the proper extension R(A) can very well be infinite, too. In such a case we can have no method for calculating R(A) in a finite number of steps. But it is still possible to talk about a decision method for R in a natural way. A relational R can be regarded as a mapping from a class of sets to a class of n-ary functions. For R is a function from $\mathbb{D}_R$ to $\mathbb{E}_R$ and an element in $\mathbb{E}_R$, for example R(A), can easily be transformed into a function, viz.

the characteristic function of R(A). We take a decision method for R to be a method by which, given an element A of $\mathbb{D}_R$, we can find a decision method for R(A) in a finite number of steps. More precisely a decision method for R is a method such that given A in $\mathbb{D}_R$ and $a_1,...,a_n$ in A we can decide in a finite number of steps whether $R(A;a_1,...,a_n)$ or not. Let C_R be the function such that

$$C_R(A,a_1,...,a_n) = \begin{array}{ll} 1 & \text{iff} \quad R(A;a_1,...,a_n) \\ 0 & \text{iff} \quad \neg R(A;a_1,...,a_n) \end{array}$$

A decision method for R is thus a decision method for C_R.

The concept of a method is of course central to the idea of a decision method. Since our concern is applied rather than pure mathematics our notion of decision method (or mechanical procedure) will be different from the one used in mathematical logic, where expressions like "ideal methods", "ideal machine" are used to signify that physical, empirical and practical limitations are ignored. If R is an empirical relational then it is reasonable that the decision method for R should be an actual empirical procedure (an experiment). A procedure in this sense is therefore not what the logicians call a mechanical method. A mechanical method is one which could be carried out by a suitably designed machine, and if we think of this machine as an industrial robot I think we come quite close to the notion of mechanical and empirical procedure that will be used here.

Let us look at a simple example (essentially taken from Krantz et.al., 1971, p. 2). Let R be the relational "longer than" with straight rigid rods as its range of application. To decide if it holds between two sticks a and b we place a and b side by side and adjust them so that one is entirely beside the other and they coincide at one end. If a then extends beyond b at the other end, then $R(\{a,b\}; a,b)$. If b extends beyond a, then $R(\{a,b\}; b,a)$. If a and b coincide at the other end, then $R(\{a,b\})$ is the empty set. Hence, the decision procedure for R* is the procedure of placing the rods pairwise beside each other so that they coincide at one end and studying which one of them extends beyond the other at the other end. If R=R* this procedure yields a decision procedure for R, too. But R(A) can also be related in a more complicated way to R*(A). Whether there is a decision method for R will then depend on how R(A) is related to R*(A).

Some much debated problems are connected with decision methods for empirical relations though the term "decision method" is seldom used in this context. The subject of the verifiability criterion of meaningfulness and the verifiability theory of truth are examples, as is the problem of meaning for empirical relations. Within the framework used here some more specific questions are the following.

Are the decision methods for interesting empirical relationals often only partial in the sense that only for certain subsets of certain domains in the range of definition could the decision method be applied? Is a decision method often a conglomerate of different empirical procedures with different but sometimes overlapping ranges of application? What cognitive (epistemological) status has the proposition that a certain procedure is a decision method for an empirical relational? Is the meaning of an empirical relational never, sometimes or always given as a decision method? Do different decision methods for a relational correspond to different interpretations of the relational?

As far as I can see it is irrelevant for what is said in this study how these questions are answered. I shall not, therefore, discuss them further. Let me end this section, however, by quoting a passage from Einstein where he discusses what is meant by simultaneity. I think that what he says is a memento for all further discussions of decision methods for empirical relations.

> The concept does not exist for the physicist until he has the possibility of discovering whether or not it is fulfilled in an actual case. We thus require a definition of simultaneity such that this definition supplies us with the method by means of which, in the present case, he can decide by experiment whether or not both the lightning strokes occurred simultaneously. As long as this requirement is not satisfied, I allow myself to be deceived as a physicist (and of course the same applies if I am not a physicist), when I imagine that I am able to attach a meaning to the statement of simultaneity. (I would ask the reader not to proceed farther until he is fully convinced on this point.) (Einstein, 1916, p. 22)

4.6 Derived and derivable relationals

In this section we shall introduce two new concepts, derived and derivable relationals and discuss to what extent derivability catches the idea of structural dependence. The notions will be used in chapter 5 in connection with social choice theory.

The use of the term "derived" in the context of relationals stems from measurement theory. Derived measurement deals mainly with the problem of introducing new scales defined in terms of some old ones. A typical example is a density scale d defined in terms of a mass scale m and a volume scale V as $d=m/V$. If d is defined in this way, then d is not a "fundamental" scale but is said to be derived from m and V. (See Roberts (1979) p. 76.) Furthermore, if d is used to define the attribute density, then density itself is in a sense derived from mass and volume. I think therefore that the notion of derived relationals introduced here catches one of the intuitive ideas behind the concept of derived measurement.

The ideas behind the concepts of derivable and derived relationals are simple enough. A relational S is *derivable* from a relational R if the following three conditions are satisfied.

1) R and S have the same range of definition $\mathbb{D}$.
2) There is a formal procedure for transforming $R(A)$ to $S(A)$ for every A in $\mathbb{D}$.
3) The transforming procedure is applicable to all possible extensions of R (and not just the actual extensions) and the result of applying the procedure to a possible extension of R is a possible extension of S.

We call such a transforming procedure a *derivation* of S from R and say that S is derivable from R *through* it. The derivation "rewrites" (or transforms) the extensions of R to extensions of S.

A relational S is *derived* from R if the following two conditions are satisfied:

(1) S is defined or introduced as follows: If A is a domain of R then $S(A)$ is the result of applying a given transformation procedure to $R(A)$.
(2) The procedure is applicable to all possible extensions of R. We say that the procedure *derives* S from R and is the *defining derivation* of S from R.

The intuitive idea is that since the defining derivation of S from R is a formal transformation procedure, then the empirical content of S is contained in the empirical content of R.

The connection between "derived" and "derivable" is intended to be the following: S is derivable from R iff there is a relational T which is derived from R such that for every A in $\mathbb{D}$ (where $\mathbb{D} = \mathbb{D}_R = \mathbb{D}_S = \mathbb{D}_T$) $S(A) = T(A)$. If we take derivations into account the connection is intended to be as follows: f is a derivation of S from R iff there is a relational T such that f is the defining derivation of T from R and for every A in $\mathbb{D}$, $S(A) = T(A)$. Hence, if S is derived from R with f as the defining derivation, then S is derivable from R and f is a derivation of S from R. Note that a derived relational is defined or constructed from another relational by means of a formal transformation procedure (a derivation). "To be derivable from" is however a relation meaningfully applied to given relationals. If S is derived from R with f as the defining derivation, then f *defines* S from R. If S is derivable from R with f as a derivation, then f *relates* S to R.

As we shall see, the translation of these rough ideas into formal definitions is not without complications. The hard thing is to make clear what a derivation, i.e. a formal procedure transforming extensions, is. Before we turn to this problem some preliminary remarks and definitions are necessary.

Relationals are functions and can therefore be terms in a functional composition. Let us look at two simple examples. If $\mathbb{D}_R = \mathbb{D}_S$ and $\mathbb{F}_{RS} = \{\langle R(A), S(A) \rangle | A \in$ $\in \mathbb{D}_S\}$ then it is trivially true that $\mathbb{F}_{RS} \circ R = S$. And if S is subordinate to R, then we have $R^S \circ R = S$, which is easy to see.

In section 3.2 we introduced the notion of an extension class. The extension class of R is the set $\{R(A) | A \in \mathbb{D}_R\}$, and we denote it by $\mathbb{E}_R$. A *possible extension* for R is an element of the characteristic class of R which has an element of $\mathbb{D}_R$ as domain. Hence, the set of possible extensions for R is

$$\{X \in \mathbb{C}_R | X \in \mathbb{D}_R\}$$

and we call it the *possible extension class* of R and denote it by $\mathbb{PE}_R$. Behind this definition lies the idea that a structure is a possible extension of R iff its domain belongs to the range of definition of R and it has the right form, i.e. is an element of $\mathbb{C}_R$.

I will now suggest that a derivation f from S to R is a function from $\mathbb{PE}_R$ into $\mathbb{PE}_S$, i.e.

$$f: \mathbb{PE}_R \to \mathbb{PE}_S$$

and it holds that

$$S = f \circ R,$$

in other words, $S(A) = f(R(A))$ for all A in $\mathbb{D}$ (where $\mathbb{D} = \mathbb{D}_R = \mathbb{D}_S$). I will further demand of f that f(A), where A in $\mathbb{PE}_R$, is a structure with A as domain. If $A = \langle A, \rho \rangle$ then we denote the relation in f(A) with f(ρ). Hence, $f(\langle A, \rho \rangle) =$ $= \langle A, f(\rho) \rangle$. As we have asserted before it is convenient to let it be clear from the context whether R(A) means

$$\langle A, \{\langle a_1, ..., a_n \rangle \in A^n | R(A; a_1, ..., a_n)\} \rangle$$

or just

$$\{\langle a_1, ..., a_n \rangle \in A^n | R(A; a_1, ..., a_n)\}.$$

Similarly, it is convenient to let it be clear from the context if $f(\langle A, \rho \rangle)$ means $\langle A, f(\rho) \rangle$ or just f(ρ).

The idea behind $\mathbb{PE}_R$ as the domain and $\mathbb{PE}_S$ as containing the codomain of f is the following: As long as A is a possible extension of R, i.e. has the right form and domain, then f will accept A as an argument and give a value which is a possible extension of S. (f has no information of the extension of R other than about their form and domain.) It is important that f is defined for $\mathbb{PE}_R$ and not just for $\mathbb{E}_R$, because we can be mistaken about what the extension of R on A is. We can wrongly believe that R(A)=A but still apply f to A. That f should be defined just for the correct values of R seems absurd. But we cannot demand that f will be applicable to what is obviously not an extension of R, i.e. not a possible extension of R. The value of f for A where A in $\mathbb{PE}_R$ but R(A)≠A does still belong to $\mathbb{PE}_S$. Otherwise f would distinguish between right and wrong statements of the form R(A)=A where A in $\mathbb{PE}_R$.

We thus demand a derivation of S from R to be a function on $\mathbb{PE}_R$ into $\mathbb{PE}_S$ such that S=f∘R and such that f(A) has A as domain; f is thus an s-function. But this is not sufficient for characterizing a derivation. To see this suppose that R and S are relationals such that $\mathbb{D}_R = \mathbb{D}_S = \mathbb{D}$ and define f for $\mathbb{PE}_R$ in the following way:

$$f(A) = S(A)$$

for all A in $\mathbb{PE}_R$. Then f is of course a function into $\mathbb{PE}_S$ and we have S(A)=f(R(A)) for all A in $\mathbb{D}$, i.e. S=f∘R. If we demand of a derivation of S from R just that it be an s-function on $\mathbb{PE}_R$ into $\mathbb{PE}_S$ then a relational S is derivable from a relational R whenever $\mathbb{D}_S = \mathbb{D}_R$. This is not what I have in mind. What we further demand of a derivation is that it is a mechanism or procedure for rewriting (transforming) R(A) to S(A) for every A in $\mathbb{D}$. We now turn to the problem of how this can be explicated. First a few examples of what we intuitively regard as derivations.

Example 1:

Suppose that R and S are binary relationals such that $\mathbb{D}_R = \mathbb{D}_S = \mathbb{D}$ and for every A in $\mathbb{D}$

$$S(A) = \{\langle x,y\rangle \in A^2 \mid \langle y,x\rangle \in R(A)\},$$

i.e. S(A) is the converse of R(A). Let f be a function on $\mathbb{PE}_R$ such that for every $\langle A,\rho\rangle$ in $\mathbb{PE}_R$

$$f(\langle A,\rho\rangle) = \{\langle x,y\rangle \in A^2 \mid \langle y,x\rangle \in \rho\} .$$

It obviously holds that

$$S(A) = f(R(A))$$

for every A in $\mathbb{D}$, so

$$S = \text{foR} .$$

Note that f is here a kind of procedure for transforming R(A) to S(A). Suppose $\langle A,\rho \rangle \in \mathbb{PE}_R$. Given a method for deciding (in a finite number of steps) for arbitrary elements $x,y \in A$ whether $\langle x,y \rangle \in \rho$ or $\langle x,y \rangle \notin \rho$, then we have a method for deciding (in a finite number of steps) if $\langle a,b \rangle \in f(\rho)$ or $\langle a,b \rangle \notin f(\rho)$: for arbitrary elements $a,b \in A$ decide whether $\langle b,a \rangle \in \rho$ or $\langle b,a \rangle \notin \rho$. If $\langle b,a \rangle \in \rho$ then $\langle a,b \rangle \in f(\rho)$. if $\langle b,a \rangle \notin \rho$ then $\langle a,b \rangle \notin f(\rho)$. Given a decision method for ρ we thus have a decision method for $f(\rho)$. I think f is a typical example of what we intuitively will count among derivations, and S is a paradigm of a derivable relational—derivable from R.

Example 2:

Suppose that R and S are binary relationals such that $\mathbb{D}_R = \mathbb{D}_S = \mathbb{D}$ and $\mathbb{D}$ consists of only finite sets. Suppose further that u is an element in U $(U = \cup \mathbb{D})$ and

$$S(A) = \{\langle x,y \rangle \in A^2 \,|\, \langle x,y \rangle \in R(A)\} \text{ if } u \in A$$
$$S(A) = \{\langle x,y \rangle \in A^2 \,|\, \langle y,x \rangle \in R(A)\} \text{ if } u \notin A .$$

Let f be a function of $\mathbb{PE}_R$ such that for every $\langle A,\rho \rangle \in \mathbb{PE}_R$

$$f(\langle A,\rho \rangle) = \{\langle x,y \rangle \in A^2 \,|\, \langle x,y \rangle \in \rho\} \text{ if } u \in A$$
$$f(\langle A,\rho \rangle) = \{\langle x,y \rangle \in A^2 \,|\, \langle y,x \rangle \in \rho\} \text{ if } u \notin A$$

Obviously S=foR. Even in this example f is a procedure for transforming R(A) to S(A). Suppose $\langle A,\rho \rangle \in \mathbb{PE}_R$. Given a method for deciding for arbitrary elements $x,y \in A$ whether $\langle x,y \rangle \in \rho$ or $\langle x,y \rangle \notin \rho$, we have a method for deciding for all $a,b \in A$ if $\langle a,b \rangle \in f(\rho)$ or not: Decide first whether $u \in A$ or $u \notin A$; it is always possible in a finite number of steps since A is finite. If $u \in A$ decide whether $\langle a,b \rangle \in \rho$ or $\langle a,b \rangle \notin \rho$; if $\langle a,b \rangle \in \rho$ then $\langle a,b \rangle \in f(\rho)$ and if $\langle a,b \rangle \notin \rho$ then $\langle a,b \rangle \notin f(\rho)$. If $u \notin A$ decide whether $\langle b,a \rangle \in \rho$ or $\langle b,a \rangle \notin \rho$; if $\langle b,a \rangle \in \rho$ then $\langle a,b \rangle \in f(\rho)$ and if $\langle b,a \rangle \notin \rho$ then $\langle a,b \rangle \notin f(\rho)$. Thus we have a method for—in a finite number of steps—finding a decision method for $f(\rho)$ given a decision method for ρ. Since S(A)=f(R(A)), we can in a finite number of steps find a decision method for S(A) given a decision method for R(A). Intuitively, f seems to be a derivation of S from R, so S is derivable from R.

.   .   .

A relational R takes sets as arguments and structures as values. A derivation f takes structures both as arguments and values. Relationals and derivations are therefore from a formal point of view quite similar. But regarding decision methods for these two kinds of objects it is convenient to distinguish two notions. In section 4.5 we discussed what a decision method for a relational would be. A decision method for R is intuitively a method by which, given an element A of $\mathbb{D}_R$, we can find a decision method for R(A) in a finite number of steps. By a decision method for f we do not mean a method by which, given an element A of $\mathbb{PE}_R$, we can find a decision method for f(A). Rather, it is sufficient that there is a method by which, given an element A of $\mathbb{PE}_R$ *and* a decision method for A, we can find a decision method for f(A) (in a finite number of steps). To be more precise, it is in fact not even necessary that there is given a *decision method* for A=⟨A,ρ⟩. The important thing is that we are furnished with an object which, when supplied with n elements $a_1,...,a_n$ in A, will tell us whether $\rho(a_1,...,a_n)$ or not. Such an object is, following Turing, often called an oracle. We thus say that a decision method for a derivation f is a method by which, given an oracle for A, we can find a decision method for f(A) (in a finite number of steps). (To decide if ⟨$a_1,...,a_n$⟩∈ f(A) the method might involve asking the oracle a finite number of questions about what holds with respect to ρ for an n-tuple of elements in A.) Let us further say that f is *relatively effective* if there is a decision method (in the above sense) for f. (I have here essentially followed Shoenfield,1967, p. 148.)

Despite their formal similarity, there is a considerable difference between relationals and derivations. Relationals might well be of an empirical character— we are here in a fact mainly interested in empirical relationals—while derivations are intended to be mathematical objects. As was pointed out in section 4.5, the decision method for a relational which is of an empirical character is not what logicians mean by a decision method. It is mechanical roughly in the sense of an industrial robot. Derivations, on the other hand, are mechanical in the sense logicians normally intend. If f is relatively effective, then the decision method for f(A), which we find in a finite number of steps—given an oracle for A—is a decision method in the logical sense of the word.

We can now define derivable and derived relationals as follows. S is *derivable* from R if $\mathbb{D}_R = \mathbb{D}_S$ and there is f: $\mathbb{PE}_R \to \mathbb{PE}_S$ such that S=foR and such that f(A) has A as its domain for all A in $\mathbb{PE}_R$ and f is relatively effective. S is *derived* from R if there is a function f with $\mathbb{PE}_R$ as domain such that S is defined as foR and f(A) has A as domain and f is relatively effective.

Note that if S is derivable from R and there is a decision method for R, then we have a decision method for S. The decision method for S consists of the decision method for R and the derivation of S from R.

Since the concept of effective function defined above is not a technically well-defined concept, this holds for derivable relations, too. But I hope that the definitions will be sufficiently clear for the use we make of the concepts, principally in chapter 5 (where I suggest that the group decision function is a derivation and the group preference relational is derivable—perhaps even derived —from the individual preference relationals).

Let us now have a look at an example of a function f which fails to have the property of relative effectiveness.

Example 3:

Suppose that R is a regional semiordering and S the weak ordering associated with R. The notion of the weak ordering associated with a semiordering was introduced in section 3.3. Hence, $\mathbb{D}_R = \mathbb{D}_S$ and we denote $\mathbb{D}_R$ simply by $\mathbb{D}$. If A is in $\mathbb{D}$ then

$$S(A) = \{\langle x,y \rangle \in A^2 \mid \forall z \in A: (R(A;z,x) \Rightarrow R(A;z,y)) \& R(A;y,z) \Rightarrow R(A;x,z)\}$$

i.e. S(A) is the weak order associated with R(A). Let f be the function defined on the class K of all semiorders such that for $\langle A,\rho \rangle$ in K

$$f(\langle A,\rho \rangle) = \langle A,\sigma \rangle$$

where σ is the weak order associated with ρ. Then it obviously holds that

$$S(A) = f(R(A))$$

for every A in $\mathbb{D}$, so

$$S = f \circ R .$$

Suppose we have an oracle for deciding, for arbitrary elements $x,y \in A$, whether $\langle x,y \rangle \in \rho$ or $\langle x,y \rangle \notin \rho$. This does not guarantee that we have a method for deciding (in a finite number of steps) whether $\langle a,b \rangle \in f(\rho)$ or $\langle a,b \rangle \notin f(\rho)$. The reason is that if A is infinite, then we might have to check whether ρ holds or does not hold for an infinite number of ordered couples to be able to decide whether $\langle a,b \rangle \in f(\rho)$ or $\langle a,b \rangle \notin f(\rho)$. For

$$\langle a,b \rangle \in f(\rho) \quad \text{iff} \quad \forall c \in A: (c\rho a \Rightarrow c\rho b) \& (b\rho c \Rightarrow a\rho c) .$$

There is no method (in the sense of mechanical) which can carry this out. f is therefore not relatively effective and if $\mathbb{D}$ contains infinite sets then S is not in general derivable from R. However, if $\mathbb{D}$ contains only finite sets then f', which is the restriction of f to the class of finite semiorders, is a derivation of S from R. f' is relatively effective since to decide whether $\langle a,b \rangle \in f'(\rho)$ or not we have only to check for a finite number of ordered couples whether they belong to ρ or not.

That S is derived from R implies of course some kind of dependence between S and R. It is natural, therefore, to ask what relation there is between derivability and subordination. At first sight one might perhaps be tempted to expect derivability to imply subordination, but the following example shows that this is not the case.

Example 4:

Suppose that R and S are strict simple orderings such that $\mathbb{D}_R = \mathbb{D}_S = \mathbb{D}$ and $\mathbb{D}$ consists only of finite sets. Suppose further that u is an element in U $(U = \cup \mathbb{D})$ and

$$S(A) = \{\langle x,y \rangle \in A^2 \,|\, \langle x,y \rangle \in R(A)\} \quad \text{if } u \in A$$
$$S(A) = \{\langle x,y \rangle \in A^2 \,|\, \langle y,x \rangle \in R(A)\} \quad \text{if } u \notin A.$$

Finally, suppose that there exist $A_1, A_2 \in \mathbb{D}$ such that $u \in A_1$, but $u \notin A_2$ and there exists $\varphi \in \mathbb{I}(R(A_1), R(A_2))$. If $\langle a,b \rangle \in R(A_1)$ then $\langle \varphi(a), \varphi(b) \rangle \in R(A_2)$ and $\langle a,b \rangle \in S(A_1)$. $\langle \varphi(a), \varphi(b) \rangle \in R(A_2)$ implies $\langle \varphi(b), \varphi(a) \rangle \in S(A_2)$. Since $S(A_2)$ is asymmetric it follows that $\langle \varphi(a), \varphi(b) \rangle \notin S(A_2)$. We thus have $\langle a,b \rangle \in S(A_1)$ but $\langle \varphi(a), \varphi(b) \rangle \notin S(A_2)$, so $\varphi \notin \mathbb{I}(S(A_1), S(A_2))$. Hence, S is not subordinate to R. But this example is a special case of example 2, which implies that S is derivable from R. Subordination does not, therefore, follow from derivability.

For R and S in example 4 it holds that
$$\mathbb{I}(S(A)) = \mathbb{I}(R(A))$$
for all $A \in \mathbb{D}$. S is therefore automorphism-preserving relative to R. (S is automorphism-preserving relative to R iff $\mathbb{I}(S(A)) \supseteq \mathbb{I}(R(A))$ for all A in $\mathbb{D}_R$.) Does it hold generally that S is automorphism-preserving relative to R if S is derivable from R? The answer is no, as the following example shows.

Example 5:

Suppose that R and S are binary relationals such that $\mathbb{D}_R = \mathbb{D}_S = \mathbb{D}$ and $\mathbb{D}$ consists of only finite sets. Suppose further that u is an element in U $(U = \cup \mathbb{D})$ and

$$S(A) = \{\langle x,y \rangle \in A^2 \,|\, [x \neq u \ \& \ y \neq u \ \& \ \langle x,y \rangle \in R(A)] \text{ or } [(x=u \text{ or } y=u) \ \& \\ \langle y,x \rangle \in R(A)]\}.$$

It is easy to see that S is derivable from R. To decide whether $\langle x,y \rangle \in S(A)$ or not ,we have to decide if any of x and y is u. If this is the case then $\langle x,y \rangle \in S(A)$ iff $\langle x,y \rangle \in R(A)$, and if this is not the case then $\langle y,x \rangle \in S(A)$ iff $\langle x,y \rangle \in R(A)$.

Suppose now that A∈ 𝔻 where A= {a,b,u} and that a and b are distinct from u. Finally suppose that

$$R(A) = \{ \langle a,b \rangle, \langle a,u \rangle, \langle u,a \rangle, \langle u,b \rangle \}$$

Then

$$S(A) = \{ \langle a,b \rangle, \langle u,a \rangle, \langle a,u \rangle, \langle b,u \rangle \}.$$

Let φ = { ⟨a,u⟩, ⟨b,b⟩, ⟨u,a⟩ }. Then φ∈𝕀(R(A)) but φ∉𝕀(S(A)). Hence, 𝕀(R(A)) is not a subset of 𝕀(S(A)). S is thus not automorphism-preserving relative to R.

Note that in example 4 and 5 the element u in U plays a special role. Let T be the unary relational such that

$$T(A;a) \text{ iff } a=u.$$

It is easy to see that in both examples 4 and 5, S is subordinate to R *and* T taken together. However, in both examples S is derivable from R alone, while in example 4 S is not subordinate to R (alone) and in example 5 S is not automorphism-preserving relative to R (alone).

In view of the above discussion, it seems reasonable to say that derivability is another form of connection between relationals, distinct from both isomorphism and automorphism preservation. The latter are invariance conditions, which is not the case for derivability. Derivability is therefore in one sense not a kind of structural dependence, but in another sense it is. We shall discuss this further in section 4.8.

Note that a derivation f of S from R contains $\mathbb{F}_{RS}$ as a subset. This implies that if f is isomorphism-preserving then this holds for $\mathbb{F}_{RS}$, too. But from the fact that $\mathbb{F}_{RS}$ is isomorphism-preserving it does not follow that f is isomorphism-preserving. Note also that ⟨A,f(A)⟩ does not necessarily belong to R^S. R^S is $\mathbb{F}_{RS}$ closed under simultaneous isomorphism, but if A∈ $\mathbb{PE}_R \backslash \mathbb{E}_R$ then there need not exist a set B in $\mathbb{D}_R$ and a bijection φ from B to A such that φ[R(B)] = A *and* φ[S(B)] = f(A).

Let us end this section with another remark on the kind of dependence that derivability implies. At the end of section 4.4 it was pointed out that subordination expresses domainwise dependence. This is also the case for derivability as we have defined it here. But the notion can be generalized so that this limitation can be overcome. We expect, for example, R* to be in general derivable, and even derived, from R in terms of these generalized notions.

4.7 Stability of transitions

The notions of domain stability and weak domain stability introduced in section 3.1 as attributes of relationals can be generalized in a natural way so that they can be applied to transitions and derivations. It might happen that the transition between two relationals is stable although the relationals themselves are not stable. Let us have a look at this.

In section 4.1 we introduced the notion of an s-function. An s-function f is a function from a class K of relational structures of the same type τ to a class L of relational structures of the same type υ such that f is domain-preserving, i.e. $f(X)$ has X as base set. Note that both a derivation of one relational from another and the univalent transition between two relationals are s-functions. We therefore formulate the notions of stability for s-functions in general.

Suppose f is an s-function on K into L. We say that f is *stable* if for all A and B in K, $A/B = B/A$ implies $f(A)/B = f(B)/A$. And we say that f is *weakly stable* if for all A and B in K, $B = A/B$ implies $f(B) = f(A)/B$.

Let us now consider stability for $\mathbb{F}_{RS}$. Suppose $\mathbb{D}_R \subseteq \mathbb{D}_S$. Then $\mathbb{F}_{RS} = \{\langle R(A),S(A)\rangle \mid A \in \mathbb{D}_R\}$ is an s-function on $\mathbb{E}_R$ into $\mathbb{E}_S$. According to the definition, $\mathbb{F}_{RS}$ is stable iff for all A and B in $\mathbb{D}_R$, $R(A)/B = R(B)/A$ implies $\mathbb{F}_{RS}(R(A))/B = \mathbb{F}_{RS}(S(B))/A$. Since $\mathbb{F}_{RS}(R(A)) = S(A)$ and $\mathbb{F}_{RS}(R(B)) = S(B)$ it holds that $\mathbb{F}_{RS}$ is stable iff for all A and B in $\mathbb{D}_R$,

$$R(A)/B = R(B)/A \;\Rightarrow\; S(A)/B = S(B)/A. \qquad (4.7.1)$$

It is easy to see now that if S is stable then $\mathbb{F}_{RS}$ is stable, too. For if S is stable then $S(A)/B = S(B)/A$ holds for all A and B in $\mathbb{D}_S$ and since $\mathbb{D}_S \supseteq \mathbb{D}_R$, (4.7.1) holds for all A and B in $\mathbb{D}_R$. Suppose now that $\mathbb{D}_R = \mathbb{D}_S$. Then it holds that if $\mathbb{F}_{RS}$ and R are stable then S is stable too, which is seen immediately from (4.7.1). However, that $\mathbb{F}_{RS}$ is stable does not alone guarantee that S is stable. If R is unstable, then S can be unstable even if $\mathbb{F}_{RS}$ is stable. We can summarize the results just mentioned in the following way; we presuppose that $\mathbb{D}_R = \mathbb{D}_S$.

$$R \text{ stable } \& \; \mathbb{F}_{RS} \text{ stable } \Rightarrow S \text{ stable}$$
$$S \text{ stable } \Rightarrow \mathbb{F}_{RS} \text{ stable}$$

Let us say that S is *stable relative to* R if $\mathbb{D}_R = \mathbb{D}_S$ and $\mathbb{F}_{RS}$ is stable. If S is stable relative to R then S is stable if R is stable. Note that if S is stable then S is stable relative to every R such that $\mathbb{D}_R = \mathbb{D}_S$.

We say that S is *weakly stable relative to* R if $\mathbb{D}_R = \mathbb{D}_S$ and $\mathbb{F}_{RS}$ is weakly stable. It is easy to see that if R is weakly stable and S is weakly stable relative to R, then S is weakly stable.

Suppose f and g are s-functions such that $g \subseteq f$. Then g is stable if f is stable. But g can of course be stable although f is not stable. The situation is analogous in the case of weak stability. Note that if f is a derivation of S from R then $\mathbb{F}_{RS} \subseteq f$. Note also that it holds generally that $\mathbb{F}_{RS} \subseteq R^S$. Hence, if a derivation of S from R stable, then $\mathbb{F}_{RS}$ is stable too. And if R^S is a function and is stable, then $\mathbb{F}_{RS}$ is stable. But it is possible that $\mathbb{F}_{RS}$ is stable but the derivation of R from S is not stable. The same holds for weak stability.

Can upward and downward stability be generalized to s-functions? The answer is yes, but the situation is somewhat complicated and the topic is postponed to Part 3 of this essay.

Let us end this section with a note on terminology. The term "stability" was first introduced in this study as an abbreviation for "domain stability", which is used of relationals. The reason for choosing this term is the following. A relational R is domain stable if the truth value of the proposition that R holds among $x_1,...,x_n$ is stable with respect to change of domain. It is not obvious how this idea should be generalized to s-functions. In any case the question is hardly one of domain stability but rather structure stability. If we express the idea of stability by the condition

$$f(A)/B = f(A/B),$$

or perhaps even better

$$A/B = A'/B \Rightarrow f(A)/B = f(A')/B,$$

then the following formulation seems not entirely unnatural. The restriction to B of a value of f for an argument A is stable with respect to change in the argument A as soon as it is always the same over B. We shall therefore apply the stability terminology in this context too. Also the analogy between the stability properties of s-functions and of relationals is a reason for this.

4.8   The structural character of transitions and subordination

It may seem obvious from what has been said already about transitions and subordination that these notions have a structural character. This section is therefore really little else than a long-winded account of what has already been stated in a more transparent and concise form in the more formal presentation.

However, the discussion in this section is perhaps not totally out of place as an attempt to emphasize the structural character of two of the main concepts in this chapter. We shall also see how the notion of concomitant variations is related to subordination.

Suppose S is subordinate to R. Then R^S is a function on $\mathbb{C}_R$ onto $\mathbb{C}_S$. There exists therefore exactly one corresponding extension of S for every possible extension of R. AR^SA' can be written $R^S(A)=A'$ and it would be tempting to interpret it as follows: A' is what the extension of S on A would be if A were the extension of R on A. Or in indicative form: the extension of S on A is A' given that the extension of R on A is A. If S is not subordinate to R, then it would be tempting to say that AR^SA' ought to be interpreted as follows: A' is a possible extension of S on A given that A is the extension of R on A. But it is something suspect with this way of interpreting R^S. To see this, suppose S is derived from R with f as the derivation. Then it is natural to regard f(A) as what the extension of A would be if the extension of R on A were A. But f can, as we saw in section 4.6, lack the property of isomorphism preservation, so we can have a case where $AR^Sf(A)$ and AR^SA', where $f(A)\neq A'$, or a case where $\neg AR^Sf(A)$. The interpretation suggested above of AR^SA' is therefore not compatible with the interpretation of f(A) as the extension of S on A given that the extension of R on A is A. A closer look at R^S is therefore necessary.

As was pointed out in section 4.1, a set-theoretical structure or system has a structure or form, and it is the structure in this latter sense which is the same (or similar) when two structures in the first sense are isomorphic. (We here use "system" for "set-theoretical structure" to avoid confusion with "structure" as form. In Gericke and Martens, 1974, the term "configuration" is used for "structure" in the sense of a system; see p. 509.) The transitions between relationals are closed under isomorphisms. Therefore, the values of R^S for a structure is determined by the form or structure of that system. AR^SA' means that it is possible for the extension of S on A to have the same structure (form) as A' *given* that R(A) has same structure (form) as A. In other words, AR^SA' implies that the structure of A' is a possible structure for the extension of S on A on the basis of the fact that the structure of R(A) is the one A has. If R^S is a function, then there is only one possible structure for the extension of S on A given that the structure of R(A) is the one A has.

We can summarize the preceding discussion in the following way. Let $\cong$ be the equivalence relation *is isomorphic to*. If R^S is a function then

$$A \cong B \text{ implies } R^S(A) \cong R^S(B).$$

Generally, when it is not given that R^S is a function, then

$$A \cong B \And AR^SA' \text{ implies } \exists B' \in C_S: BR^SB' \And A' \cong B'.$$

If R^S is a function then the structure of $R(A)$ determines completely the structure of $S(A)$.

However, this is not the whole story. If the structure of $R(A)$ determines the structure of $S(A)$ (in the sense described above), then $S(A)$ would be determined up to isomorphism. But as we have seen, $S(A)$ is completely determined by $R(A)$ when R^S is a function. Consequently, something is missing and we must look for it.

Let $\cong$ be the equivalence relation *is isomorphic to* as before. Construct $C_R/\cong$ and $C_S/\cong$, i.e. C_R modulo $\cong$ and C_S modulo $\cong$. An element in $C_R/\cong$ is thus a set consisting of all structures in C_R, which are isomorphic to each other. Define $[R^S]$ as a subset of the Cartesian product of $C_R/\cong$ and $C_S/\cong$ by

$$[A][R^S][A'] \text{ iff } AR^SA'$$

where $[X]$ is the equivalence class modulo $\cong$ generated by X. $[R^S]$ may be regarded as the correspondence of R^S which relates structures (forms) of systems. However, $[R^S]$ is not well-defined. For $[R^S]$ to be well-defined it must hold that

$$A \cong B \And A \cong B' \And AR^SA' \text{ implies } BR^SB'.$$

But obviously this does not hold. Instead the following holds:

$$\phi \in \mathbb{I}(A,B) \And \phi \in \mathbb{I}(A',B') \And AR^SA' \text{ implies } BR^SB'.$$

What is important to remember is that R^S is $\mathbb{F}_{RS}$ closed under isomorphisms *simultaneously* for both components.

It is important here to be clear on what we mean by saying that two systems are isomorphic. $X = \langle X, \rho \rangle$ and $Y = \langle Y, \sigma \rangle$ (which for simplicity we think of as binary relations) are isomorphic iff there is a way of identifying each element in X with a distinct element in Y (i.e. a bijection ϕ from X to Y) and ρ holds between two elements in X iff σ holds between the corresponding elements in Y (i.e. $x\rho y$ iff $\phi(x)\sigma\phi(y)$). It is relative to a way (rule) of identifying the elements in X with those in Y, that we can say that X and Y have the same structure. Note that if two systems X and Y have the same structure (form) when we identify the elements in X with those in Y according to the bijection ϕ, then $\phi[X] = Y$, i.e. $\phi[X]$ and Y are the same structure (system). It is obvious that X and Y may have the same structure relative to one rule of identification but not relative to another.

Since R^S is $\mathbb{F}_{RS}$ closed under isomorphism *simultaneously* for both components, $\langle X, Y \rangle$, where $X = Y$, belongs to R^S iff there is an A in $\mathbb{D}_R$ such that the structure of **X** is the same as the structure of R(A) and the structure of **Y** is the same as the structure of S(A), in both cases relative to the same way of identifying the elements in A with those in **X**. Thus, whenever R^S is a function then the following holds: If the structure of R(A) is the same as the structure of R(B) when we identify the elements in A and B according to φ, then the structure of S(A) is the same as the structure of S(B) when φ identifies the elements. Or, more formally,

$$\varphi \in \mathbb{I}(R(A), R(B)) \;\Rightarrow\; \varphi \in \mathbb{I}(S(A), S(B)).$$

This can be equivalently formulated:

$$\varphi[R(A)] = R(B) \;\Rightarrow\; \varphi[S(A)] = S(B).$$

Now it is a small step to the following principle. If R^S is a function, then **BR^{S}B'** means that if the structure of R(A) is the same as that of **B** with φ as the identification of the elements in A with those in **B**, then the structure of S(A) is the same as that of **B'** with φ as the identification of the elements. In other words,

$$\textbf{BR}^S\textbf{B'} \;\&\; \varphi[R(A)] = \textbf{B} \quad \text{implies} \quad \varphi[S(A)] = \textbf{B'}$$

which is easily seen to be true. We can therefore say that when R^S is a function, then the structure of R(A) determines the structure of S(A) with the same identification of the elements. We might also say that when R^S is a function, then the structure of R(A) determines S(A) and S(A) is structurally determined by R(A).

If we say that **X** has the same structure as **Y** and $X = Y$, we often presuppose that the identification of the elements in X with those in Y is the identity mapping ι_X. In that case $\iota_X[X] = Y$, i.e. $X = Y$. Therefore, if we say that the structure of the extension of R on A is the same as the structure of **B** where $B = A$, it is often natural to interpret this as R(A) = **B**. In such a case, **AR^{S}A'** means, when R^S is a function, that if R(A) = **A** then S(A) = **A'**; or in other words:

the extension of S on A is A' given that the extension of R on A is A (*)

which is the interpretation of **AR^{S}A'** that we started with and rejected. However, we now see that it is correct as long as we remember its hidden "structural" character. We might perhaps say that (*) holds for *structural reasons*.

Suppose now that R^S is not known to be a function. What, then, does **BR^{S}B'** mean? Note first that **BR^{S}B'** iff there is a C in $\mathbb{D}_R$ such that the structure of R(C) is the same as the structure of **B** with ψ as the identification and S(C) has the same structure as **B'** with ψ as the identification. Or more formally,

$$\exists \; C \in \mathbb{D}_R: \psi[R(C)] = \mathbf{B} \; \& \; \psi[S(C)] = \mathbf{B'}$$

(which is easily seen to be true). $\mathbf{B}$ as the structure of an extension of R and $\mathbf{B'}$ as the structure of an extension of S are thus simultaneously realizable. It is therefore structurally possible that the structure of S(A) with φ as the identification of the elements is $\mathbf{B'}$, given that the structure of R(A) with φ as the identification of the elements is $\mathbf{B}$. Hiding the role of the structure (form) in the same way as in (*) above, we can say that AR^SA' means that A' is a possible extension of S on A given that A is the extension of R on A. But we might perhaps say that this holds for structural reasons.

From this rather long-winded discussion the structural character of R^S is hopefully clear. We use R^S to define a kind of dependence, viz. subordination, which holds between S and R when R^S is a function (i.e. univalent) and this is of course a dependence on structural grounds—a kind of structural dependence. S is subordinate to R when for all A in $\mathbb{D}$, (where $\mathbb{D} = \mathbb{D}_R = \mathbb{D}_S$) S(A) is (completely) structurally determined by R(A).

Let us now try another line of attack. As was pointed out in chapter 1, the notion of dependence we are interested in is intimately associated with the idea of concomitant variation. The basic conception is that two variables are dependent if the change in one variable is accompanied by a change in the other. We shall now try to apply this idea to relationals.

The extension of R on A states what holds for the elements in A with respect to R. Suppose that the extension of R on A is $\mathbf{A}$ and that the extension of S on A is $\mathbf{A'}$. Suppose now that we change the extension of R on A to $\mathbf{A}_1$, where $\mathbf{A}_1 = \mathbf{A}$, what will then happen to the extension of S on A?

The first problem here is what it means to change the extension of R on A from A to $\mathbf{A}_1$. A change in a concrete or "physical" sense will often involve a change of the elements in A and thereby a change of A, and this is exactly what we want to avoid. The change of R(A) is hypothetical and our problem is therefore best expressed by a counterfactual: What would the extension of S on A be if the extension of R on A were $\mathbf{A}_1$? The following is a natural idea. Look for a set B such that if we identify the elements in A with those in B according to φ (φ is a bijection from A to B), then the same holds for the elements in A—when the extension of R on A is $\mathbf{A}_1$—as that which holds for the elements in B with respect to R. In other words, the structure of $\mathbf{A}_1$ and R(B) is the same when φ identifies the elements in A and B. Thus,

$$\varphi[A_1] = R(B) .$$

When such a B has been found, determine the extension of S on B, i.e. S(B). Then find A_1' such that the same holds for the elements in A (A=A_1') when the extension of S on A is A_1' as that which holds for the elements in B when the extension of S on B is S(B), *given* that φ identifies the elements in A and B. In other words, find A_1' such that the structure of A_1' and S(B) is the same with φ as the identification of A with B. Then,

$$\varphi[A_1'] = S(B).$$

This implies that

$$A_1' = \varphi^{-1}[S(B)] .$$

A_1' is one of the things the extension of S on A could be if the extension of R on A were A_1, and we can regard A_1' as one possible extension of S on A given that the extension of S on A is A_1' , i.e. A_1' is a possible value for S(A) when R(A)=A_1'. The set of all such possible extensions is

$$\{\varphi^{-1}[S(B)] \mid B \in \mathbb{D} \text{ and } \varphi[A_1] = R(B)\}.$$

It is rather easy to see (the proof is postponed until section 7.2) that this set is the same as $R^S[A_1']$ and consists therefore of exactly one element for every A_1' iff R^S is a function. This implies that we can for all A in $\mathbb{D}_R$ say unambiguously what the extension of S on A would be if the extension of R on A were A_1' iff R^S is a function. The variation of S is therefore completely determined by the variation of R iff R^S is a function, i.e. iff S is subordinate to R. And R^S expresses, in the sense described above, the possible variations of S when R varies.

It is at this point important to note the following. S is subordinate to R iff a change in what holds for the elements of A with respect to S is always accompanied by a change in what holds for the elements of A with respect to R. But not the other way round, which it might be tempting to believe; i.e. it is not the case that change in what holds for the elements of A with respect to R is always accompanied by a change in what holds for the elements of A with respect to S.. For S is subordinate to R iff R^S is a function, and this implies that if the extension of S on A changes then the extension of R on A must also change. Thus, if R^S is a function then a variation in S must always be accompanied by a variation in R, but a variation in R need not be accompanied by a variation in S. However, the variation in R is not completely determined by the variation in S.

We have of course interpreted variation here in a purely structural manner.

The change of the extension of R on A from A to A_1 has been regarded as a change in the structure of the extension of R on A. And to determine what happens with the extension of S on A when the extension of R on A changes from A to A_1, we look for a B such that the structure of A_1 is the same as the structure of S(B) with φ as the identification, i.e. $\varphi[A_1] = R(B)$. There is, however, at least one alternative to this procedure. We can describe a change in the extension of R on A by identifying each element a in A with an element $\varphi(a)$ such that after the change a is, with respect to R, indistinguishable from $\varphi(a)$. We then study what holds with respect to S for all the $\varphi(a)$s, and apply the principle that for all a in A, $\varphi(a)$ and a is indistinguishable with respect to S . What then holds of all a in A with respect to S is what holds of the elements in A after the change. To develop this idea more formally some technical notions which have not so far been presented are needed, and we are not going to pursue this idea further in this essay.

We end this section with a few remarks on derivations and derivability.

If S is derived from R with f as the derivation then note the difference between f(A)=A' and AR^SA'. f(A)=A' means that the extension of S on A is A', given that the extension of R on A is A. If R^S is a function then $R^S(A)=A'$ means the same but on the basis of the structure of the extension of R on A. If R^S is not known to be a function, then AR^SA' means that A' is a possible extension of S on A on the basis of the structure of the extension of R on A given that R(A)=A. If f is isomorphism-preserving, then $f(A)=R^S(A)$ for all $A \in \mathbb{PE}_R$. f(A) is then determined on the basis of the structure of A, which is just what we would expect if f is isomorphism-preserving. The difference between f and R^S thus becomes clear: For f systems are relevant while for R^S only the structure of the systems has relevance. R^S "operates" on structures but f in the general case does not.

As we have seen, a derivation does not generally "operate" on structures and that S is derived from R is hardly a structural connection, i.e. hardly expresses a structural dependence between S and R. R is rather in some sense prior to S. However, if S is derivable from R I think that the dependence between S and R is at least in a weak sense a structural one. The relation between S and R is then purely formal and involves neither causality nor priority of any kind.

If S is derived from R with f as the derivation, then the counterfactual "the extension of S on A would be A' if the extension of R on A were A"", where A is in $\mathbb{PE}_R$ but R(A)≠A , has a truth-value. It is true iff f(A)=A'. But if S is derivable from R with f as a derivation while S is not derived from R, then it is doubtful whether the above counterfactual has a truth value. Compare this with what holds for subordination. Should S be subordinate to R, we can say that the counterfactual is true iff $R^S(A)=A'$. But if S is not subordinate to R it has a truth value iff A' is the only structure that is related to A by R^S; note that this may hold for some A even if R^S is not a function.

The idea with which we began the discussion of derivability in section 4.6 was roughly the following: S is derivable from R if there is a formal procedure for transforming R(A) into S(A) for every A in $\mathbb{D}$. We took the procedure to be a function f on $\mathbb{PE}_R$ into $\mathbb{PE}_S$ such that S=foR and such that f(A) has A as domain. The point of the condition that the procedure be formal is that the empirical content of S be contained in the empirical content of R. The procedure will then not contribute any empirical information itself. f must therefore be a formal or mathematical function. But how should this vague notion be explicated? The line of thought pursued in section 4.6 was that f is an effective function (relatively effective, to be more precise). This seems to guarantee that f is formal or mathematical. It also emphasizes that f is a procedure or method. The notion of derivability we have arrived at is in a sense a strong one. This can be seen from example 3 in section 4.6. If R is a regional semiordering and S the weak ordering associated with R and f is defined such that S=foR, then it is not certain that f is relatively effective. But f is of course a mathematical function.

It seems to me that there are two competing ideas of what derivability means, one weaker and one stronger. According to the weaker idea a derivation is a mathematical non-empirical function, while according to the stronger a derivation is a mechanical method or procedure. In section 4.6 I seized upon the stronger notion. How can the weaker notion be explicated? The function associating a weak order to a semiorder is a mathematical function, perhaps because it operates only on structures, i.e. is isomorphism-preserving. This leads to the following conjecture. A derivation (in the weaker) sense is either relatively effective or isomorphism-preserving. But I think that even this is too narrow. A derivation may consist of different rules each one either relatively effective or isomorphism-preserving *and* a method for deciding which rule should be applied for a given argument. Somewhat more precisely, f is a derivation of S from R if S is defined as S=foR, where f is a function on $\mathbb{PE}_R$ into $\mathbb{PE}_S$ and f(A) has A as domain, and there is a partition $\wp$ of $\mathbb{PE}_R$ and for every $\mathfrak{R} \in \wp$ there is a relatively effective or isomorphism-preserving function $f_\mathfrak{R}$ such that if $A \in \mathfrak{R}$ then $f(A)=f_\mathfrak{R}(A)$ and there is an effective method, given **A**, for finding $\mathfrak{R}$ such that $\mathbf{A} \in \mathfrak{R} \in \wp$ (and thereby for finding $f_{\mathfrak{R}}$ such that $f(\mathbf{A}) = f_{\mathfrak{R}}(\mathbf{A})$).

I shall not investigate this line of thought any further in this essay, and understand derivability in the sequel to mean derivability in the stronger sense, as in section 4.6.

4.9 Significance

It seems to me that subordination catches at least one idea of complete determination within the context of relationals. And uncorrelation appears to be the appropriate notion of complete undetermination corresponding to subordination. We shall now turn to the question of how the idea of relevance can be made precise within the framework of relationals when complete determination is understood as subordination.

In this section we study the subordination of a relational S to a system of relationals $\langle Q,R \rangle$ and the transition from $\langle Q,R \rangle$ to S. We must therefore generalize the notion of subordination and transition somewhat, but this is straightforward. Let Q,R and S have $\mathbb{D}$ as range of definition. We say that S is *subordinate to* $\langle Q,R \rangle$ if for all A and B in $\mathbb{D}$,

$$\mathbb{I}(\langle Q,R \rangle(A), \langle Q,R \rangle(B)) \subseteq \mathbb{I}(S(A), S(B)) .$$

The transition from $\langle Q,R \rangle$ to S is

$\{\langle \mathbf{X},\mathbf{Y} \rangle \in \mathbb{C}_{\langle Q,R \rangle} \times \mathbb{C}_S \mid \mathbf{X} = \mathbf{Y}$ & $\exists \mathbf{A} \in \mathbb{D}: \exists \varphi \in \mathbb{B}i(\mathbf{X},\mathbf{A}): \varphi \in \mathbb{I}(\mathbf{X}, \langle Q,R \rangle(\mathbf{A}))$ & $\varphi \in \mathbb{I}(\mathbf{Y}, S(\mathbf{A}))\}$ and is denoted by $\langle Q,R \rangle^S$.

Let us now consider the following two situations.

(1) S is subordinate to $\langle Q,R \rangle$ but S is not subordinate to Q. Then R is significant for the subordination of S to $\langle Q,R \rangle$.

(2) S is correlated with $\langle Q,R \rangle$ but S is uncorrelated with Q. Then R is significant for the correlation of S with $\langle Q,R \rangle$.

In both (1) and (2) above, it seems reasonable to say that R is relevant for S in $\langle Q,R \rangle$. We thus relativize the notion of R being relevant for S to a particular system, in this case $\langle Q,R \rangle$. In (1) R is significant for the subordination of S to $\langle Q,R \rangle$, while in (2) R is significant for the correlation of S with $\langle Q,R \rangle$. Note that in (1) R is perhaps not significant for the correlation of S with $\langle Q,R \rangle$; it might hold that S is correlated with Q although S is not subordinate to Q. A rather strong notion of relevance is described in the following situation.

(3) S is subordinate to $\langle Q,R \rangle$ but S is uncorrelated with Q.

If $\mathbb{C}_S$ is non-degenerate (see section 4.3 or 7.3 for a definition) and (3) holds, then R is significant both for the subordination of S to $\langle Q,R \rangle$ and for the correlation of S with $\langle Q,R \rangle$.

This remark motivates the following definitions. Let us say that R is *s-significant* for S with respect to Q if S is subordinate to $\langle Q,R \rangle$ and S is not subordinate to Q. Let us further say that R is *c-significant* for S with respect to Q if S is correlated with $\langle Q,R \rangle$ and S is uncorrelated with Q. Finally, we say that R is *strongly significant* for S with respect to Q if S is subordinate to $\langle Q,R \rangle$ but S is uncorrelated with Q.

If R is not s-significant for S with respect to Q then we say that R is *s-insignificant* for S with respect to Q. Analogously, if R is not c-significant for S with respect to Q, then we say that R is *c-insignificant* for S with respect to Q. Thus, R is s-insignificant for S with respect to Q iff S is not subordinate to $\langle Q,R \rangle$ or S is subordinate to Q. And R is c-insignificant for S with respect to Q iff S is uncorrelated with $\langle Q,R \rangle$ or S is correlated with Q.

It is easy to see that it holds generally that R is subordinate to $\langle Q,R \rangle$. (This is proved in section 7.4.) Suppose now that Q is uncorrelated with R. Then it holds that R is strongly significant for R with respect to Q.

The transition $\langle Q,R \rangle^S$ represents in a sense the dependence or connection between S and $\langle Q,R \rangle$. It is therefore reasonable to use $\langle Q,R \rangle^S$ to define a more general notion of significance than the notions above. Let us say that R is *significant* for S with respect to Q if there are $X = \langle X,\pi \rangle \in \mathbb{C}_Q$, $Y = \langle X,\rho \rangle \in \mathbb{C}_R$ and $Z \in \mathbb{C}_S$ such that XQ^RY and XQ^SZ and $\neg(\langle X,\pi,\rho \rangle \langle Q,R \rangle^S Z)$. We shall study this notion in section 7.5. The basic idea behind this definition is that R is relevant for S with respect to Q if, at least for some domain A, the set of possible extensions of S over A is more restricted by the extension of Q *and* R over A than by the extension of Q over A alone. (That this is an interpretation of the definition is especially clear from what is said in section 7.5.) It is easy to show that if R is s-significant or c-significant or strongly significant for S with respect to Q, then R is significant for S with respect to Q. It also holds that if R is significant for S with respect to Q and S is subordinate to $\langle Q,R \rangle$ then R is s-significant for S with respect to Q. However, if R is significant for S with respect to Q and S is correlated with $\langle Q,R \rangle$, then it does not follow that R is c-significant for S with respect to Q.

In chapter 1 it was maintained that dependence could mean either determination or relevance. That β is dependent on α can therefore be explicated both as "β is determined by α" and as "α is relevant for β". Note the order between α and β in the explication of dependence in terms of relevance: "β is dependent on α" means that α is relevant to β. We have introduced *significance* as a technical term for relevance in a special context. That R is significant for S is thus one explication of S being dependent on R. It might be convenient to have a term for the converse of "being significant for". The terms "conditioned on" and "contingent on" suggest themselves.

CHAPTER 5

AN EXAMPLE: SOCIAL CHOICE

5.0 Introduction

In this chapter we shall apply some of the concepts presented in earlier chapters to a well-known subject, social choice theory. Social choice theory is rather closely related to group decision theory, although the exact character of the relation is hard to state. What is said here seems equally applicable to both subjects. We use the terminology from social choice theory essentially in accordance with Arrow (1951, 1963), Sen (1970) and Luce & Raiffa (1957). Our purpose is primarily to exemplify what has been said in earlier sections and not to conduct an analysis of the problem of social choice.

In section 5.1 we sketch the idea of dependence between individual and social preference relations, which seems to be fundamental in the theory of social choice. We make a slight transformation of social choice theory in section 5.2 by regarding the individual and social preference relations as relationals. In section 5.3 we apply the notions of isomorphism preservation and subordination to collective choice rules and to the connection between social and individual preference relations. In section 5.4 we discuss whether a collective choice rule ought to be relatively effective and whether social preferences are derivable from individual preferences. We also take a closer look at the notion "to be a function of" in section 5.3 and 5.4. We apply the concept of stability to social choice theory in section 5.5 and introduce the idea of backgrounds. In section 5.6 we say a few words about the theory of aggregation.

5.1 The notion of dependence in social choice theory

A central question in the theory of social choice is the relationship between individual preference relations and the social preference relation. The first paragraph of section 1.2 in Sen (1970) runs as follows:

> To assert that social choices should depend on individual preferences leaves the question open as to what should be the form in which individual preferences would be relevant. In his classic study, Arrow (1951) takes orderings of the individuals over the set of alternative social states to be the basic constituent of collective choice. He is concerned with rules of collective choice which make the preference ordering of the society a function of individual preference orderings, so that if the latter set is specified, the former must be fully determined. (Sen, 1970, p. 2)

In the above quotation from Sen where he describes Arrow's work he talks about the preference ordering of the society as a function of individual preference orderings, "so that if the latter set is specified, the former must be fully determined". Arrow himself puts it this way:

> ... for each pair of social states the choice depends on the ordering relations of all individuals, i.e., depends on $R_1,...,R_n$, where n is the number of individuals in the community. Put otherwise, the whole social ordering R is to be determined by the individual ordering relations for social states, $R_1,...,R_n$. We do not exclude here the possibility that some or all of the choices between pairs of social states made by society might be independent of the preferences of certain particular individuals, just as a function of several variables might be independent of some of them. (Arrow, 1963, p. 23)

Note how Sen and Arrow use the expressions "dependent on", "relevant", "a function of", "fully determined", determined by" and "independent of". Our concern in this chapter is how these phrases might be explicated when social choice theory is formulated in terms of relationals.

The kind of dependence and relationship between the social preference relation and the individual preference relations Arrow and Sen have in mind, i.e. that the social preference ordering is determined by—and a function of—the individual preference orderings, is usually explicated in the following way. There is a function, often called a social welfare function or a collective choice rule, which takes the individual's preference relations as arguments and social preference relations as values. Let us look at this idea more carefully.

In Sen (1970) section 2*1, the notion of a collective choice rule is introduced in the following way:

> Let X be the set of social states. The preference relation of the ith individual is R_i, and let there be n such persons, i=1,...,n. Let R refer to the social preference relation. ...
> Definition 2*1. A collective choice rule is a functional relation f such that for any set of n individual orderings $R_1,...,R_n$ (one ordering for each individual), one and only one social

preference relation R is determined, $R=f(R_1,...,R_n)$. (Sen, 1970, p. 28)

Sen distinguishes between preference relations and preference orderings. A preference ordering is a weak order (see Sen, 1970, p. 28). In the definition of the collective choice rule the individual preference relations are preference orderings but this is not demanded of the social preference relations. We follow this convention here and suppose that individual preference relations are weak orders but that the social preference relation need not be a weak order.

By a collective choice rule is meant a function f. It is important here to make clear what the arguments and values of a collective choice rule are and this may be done using the concept "preference profile". (See Luce & Raiffa, 1957, p. 332 and Fishburn, 1973, p. 15).

Let X be the set of social states and Π the set of all possible individual preference relations on X, and, finally, Ω the set of all possible social preference relations on X. By a preference profile for the individuals of the society is meant an n-tuple $\langle R_1,...,R_n \rangle$ of relations where R_i is the preference ordering for the ith individual. The set of possible preference profiles is thus $\Pi \times ... \times \Pi$ (where Π occurs n times) and will be denoted $\Pi^{(n)}$. According to the assumption above about the formal properties of preference relations, Π is the set of weak orders on X and Ω a set of binary relations on X.

A collective choice rule f is a function from the set of all possible preference profiles, $\Pi^{(n)}$, into the set Ω of all possible binary orderings on X, i.e.
$$f: \Pi^{(n)} \to \Omega .$$
The arguments of f are thus possible preference profiles and the values are possible preference relations. (I have essentially followed Luce & Raiffa, 1957, p. 332 here.)

As is pointed out in Luce & Raiffa (1957) p. 331, a collective choice rule can be interpreted as a procedure for passing from the individuals' preferences to the social preference. That there is a collective choice rule relating preference profiles and social preference relations is often regarded as an explication of the idea that the social preference relation is determined by—and a function of—the individual preference orderings.

In social choice theory one is interested in the conditions a collective choice rule must satisfy to be an adequate procedure for passing from the individual preference relations to the social preference relation. As Luce & Raiffa point out (p. 329), most collective choice rules are "illfare" rather than "welfare" functions, which raises the problem "what can one intuitively mean by 'welfare' function?" (See Luce & Raiffa, 1957, p. 329.) We will not deal here with that problem but restrict ourselves to the question of the relationship between the individual and social preferences.

5.2 Preference relationals and collective choice rules

It is clear that a "preference relation" or a "preference ordering" in social choice theory is a relation in the usual set-theoretical sense. But in the context of this study it also seems natural to regard them as functions, i.e. as relationals. Let us do so and see where it leads us.

R_i is the preference relational for the ith individual and R is the social preference relational. Let $\mathbb{D}$ be the range of definition of the social preference relational, and suppose for simplicity that it is also the range of definition for every individual relational. For every element A in $\mathbb{D}$, $R_i(A)$ is the extension on A of the individual i's preference relational and R(A) is the extension on A of the social preference relational. We suppose that all the individual preference relationals are weak orderings. The social preference relational need not, however, be a weak ordering. (This is in accordance with the assumption in the previous section that the individual preference relation—but not the social preference relation—in the usual interpretation as a set-theoretical relation is a weak order.) The characteristic class of individual i's preference relational is denoted by $\mathbb{C}_{R_i}$. The characteristic class of the social preference relation will be denoted by $\mathbb{C}_R$. For convenience, we now make the following assumption. If A is an element in $\mathbb{D}$, then every weak order on A is an element in $\mathbb{C}_{R_i}$ for every i, and every binary relation on A is an element in $\mathbb{C}_R$.

Let us look at the role of Π, $\Pi^{(n)}$ and Ω for preference relationals. First a convention. For an arbitrary relational Q, and any element A in $\mathbb{D}_Q$, let $\mathbb{PE}_Q | A = \{Y \in \mathbb{C}_Q | Y = A\}$. $\mathbb{PE}_Q | A$ is thus the set of possible extensions of Q on A. In this context it is convenient to let $\mathbb{PE}_Q$ denote the set of possible *proper* extensions. Thus, $\mathbb{PE}_Q | A = \{\rho | \langle A, \rho \rangle \in \mathbb{C}_Q\}$. Note now that Π is the set of possible extensions for any of the individual preference relationals $R_1,...,R_n$ on X , i.e.

$$\Pi = \{\rho | \langle X, \rho \rangle \in \mathbb{C}_{R_i}\} = \mathbb{PE}_{R_i} | X .$$

And $\Pi^{(n)}$, the set of possible preference profiles, is the set of possible extensions for $\langle R_1,..,R_n \rangle$ on X, i.e.

$$\Pi^{(n)} = \{\langle \rho_1,...,\rho_n \rangle \mid \langle X,\rho_1,...,\rho_n \rangle \in C_{\langle R_1,...,R_n \rangle}\} = \mathbb{PE}_{\langle R_1,...,R_n \rangle} \mid X.$$

According to the assumption about the formal properties of the individual prefer- ence relationals, $\mathbb{PE}_{R_i} \mid X$ is the set of weak orders on X and $\mathbb{PE}_{\langle R_1,...,R_n \rangle} \mid X$ is the set of k-tuples of weak orders on X. A possible preference profile $\langle \rho_1,...,\rho_n \rangle$ thus consists of one possible extension for each of the individual preference rela- tionals. More precisely, ρ_i is a possible extension for R_i on X. Ω, finally, is the set of possible extensions for the social preference relational on X, i.e.

$$\Omega = \{\rho \mid \langle X,\rho \rangle \in C_R\} = \mathbb{PE}_R \mid X.$$

A collective choice rule f has the set of possible preference profiles as its domain and Ω as its range. In other words,

$$f: \Pi^{(n)} \to \Omega$$

or equivalently

$$f: \mathbb{PE}_{\langle R_1,...,R_n \rangle} \mid X \to \mathbb{PE}_R \mid X.$$

We can regard a collective choice rule f as determining the extension on X of a preference relational, *given* the extension of the individual preference relationals on X. If the extension of the ith individual's preference relational on X is ρ_i, then the extension of the preference relational f determines is $f(\rho_1,...,\rho_n)$.

As was said before, a collective choice rule f can be interpreted as a procedure for passing from the individuals' preference relationals to the social preference relational. $f(\rho_1,...,\rho_n)=\rho$ means that according to f the extension of R on X would be ρ if for all i, the extension of R_i on X were ρ_i. In other words, $f(\rho_1,...,\rho_n)=\rho$ means that $R(X)=\rho$ given that $R_i(X)=\rho_i$ for all i—N.B., if f is the procedure for passing from the individuals' preference relationals on X to the social preference relational on X. Thus f determines for every possible extension on X of the individuals' preference relationals an ordering on X. This ordering is according to the procedure f the social preference relational on X given the extension on X of the individuals' preference relationals.

It is natural at this point to make one adjustment to the concept of a collective choice rule. The role of X in the presentation of social choice theory we have followed hitherto is not as clear as it ought to be. X could hardly be the set of social states under consideration in a situation because a collective choice rule (regarded as a procedure) is not restricted to the set of alternatives in a special situation but should be applicable to different situations with a different set of social states under consideration. Therefore collective choice rules will not be defined for preference profiles over X but for preference profiles over elements in $\mathbb{D}$. Thus, we let $\Pi^{(n)}$ be the set of possible extensions for $\langle R_1,...,R_n \rangle$. (We can, perhaps, think of X= U, so $\mathbb{D}$ is a class of non-empty subsets of X. In most contexts this is not appropriate, however. We shall discuss this further in the section on backgrounds, 5.5.) From now on we will also use the convention of considering graphs rather than proper extensions, and accordingly denote an order by $\langle A,\rho \rangle$ and not simply ρ (where ρ is an order on A). $\Pi^{(n)}$ is thus the set

$$\{\langle A,\rho_1,...,\rho_n \rangle \mid A \in \mathbb{D} \text{ and } \langle A,\rho_i \rangle \text{ are weak orders for all i, } 1 \leq i \leq n\}.$$

And a collective choice rule is a function of this set into

$$\{\langle A,\rho \rangle \mid A \in \mathbb{D} \text{ and } \langle A,\rho \rangle \text{ is a binary relation}\},$$

i.e.

$$f: \mathbb{PE}_{\langle R_1,...,R_n \rangle} \to \mathbb{PE}_R.$$

Note that the base set of an argument for a collective choice rule f and the base set of its values always is the same set; i.e. if f($\underline{X}$)=$\underline{Y}$ then X=Y. A collective choice rule is thus domain-preserving and is therefore an s-function. If a collective choice rule f is the procedure for passing from $R_1,...,R_n$ to R then

$$R(A) = f(\langle A,R_1(A),...,R_n(A) \rangle)$$

for all $A \in \mathbb{D}$. Consequently,

$$R = f \circ \langle R_1,...,R_n \rangle.$$

And as we have just seen, f is a function defined on $\mathbb{PE}_{\langle R_1,...,R_n \rangle}$.

Let us at this point introduce a new term. We say that a relational Q is *fixed* to a relational P *by* the function f—and that f *fixes Q to P*—if

$$\mathbb{D}_Q = \mathbb{D}_P \ \& \ f: \mathbb{PE}_P \to \mathbb{PE}_Q \ \& \ Q = f \circ P.$$

The generalization of this definition to finite relational systems is straightforward.

That Q is fixed to P by f is a necessary condition for f being a derivation of Q from P. But as we saw in section 4.6 it is not sufficient. In fact, for every pair of relationals P and Q with the same range of definition, Q can always be fixed to P. For f: $\mathbb{PE}_P \to \mathbb{PE}_Q$ such that f($\underline{A}$)=Q(A) for all $\underline{A}$ in $\mathbb{PE}_P$ fixes Q to P. That one relational is fixed to another (by some function f) says nothing about the relationship between them (besides their having the same range of definition). However, that a certain f fixes a relational Q to another relational P can be informative.

We return now to social choice theory. Suppose R is fixed to $\langle R_1,...,R_n \rangle$ by the collective choice rule f. This raises two questions. Let us start with the following. Is R= f $\circ$ $\langle R_1,...,R_n \rangle$ a kind of definition of R from $\langle R_1,...,R_n \rangle$, so that R is introduced as f $\circ$ $\langle R_1,...,R_n \rangle$? Or does R= f $\circ$ $\langle R_1,...,R_n \rangle$ just express a connection between R and $\langle R_1,...,R_n \rangle$? The difference between these two kinds of possibilities is analogous to the difference between derived and derivable relationals. It appears to me that social choice theory is a rather heterogeneous theory formation, so both possibilities may be realized but in different contexts. It seems easy to find examples of contexts where R is introduced as fo $\langle R_1,...,R_n \rangle$. The idea that a collective choice rule can be interpreted as a procedure for passing from the individuals' preferences to the social preference is in accordance with R being defined as fo $\langle R_1,...,R_n \rangle$. If R= fo $\langle R_1,...,R_n \rangle$ establishes a connection between R, already introduced, and $R_1,...,R_n$, then R can of course have been defined by R= go $\langle R_1,...,R_n \rangle$, i.e. as fixed to $\langle R_1,...,R_n \rangle$ by another collective choice rule g. Whether the meaning of R can be given in some other way is a problem we shall not enter into here. The situation is complicated since, among other things, it seems not to be uncommon in the literature to oscillate between (1) R as *the* social preference relational and then look for the adequate collective choice rule which determines R and (2) R as the social preference relational *according to* f and thus R as *the* social preference relational when f is the adequate collective choice rule.

Within the traditional view of social choice theory, where the preference relations are sets of ordered pairs, the existence of a collective choice rule relating preference profiles and social preference relations is often regarded as an explication of the idea that the social preference relation is determined by—and a function of—the individual preference orderings. When we move to a framework where preferences are expressed in terms of relationals, then it could be tempting to think that the idea that R is determined by or wholly dependent on $\langle R_1,...,R_n \rangle$ might be explicated analogously in the following way: There is a collective choice rule which fixes R to $\langle R_1,...,R_n \rangle$. But if two relationals have the same range of definition, then each is fixed to the other by some function. There is therefore always a function which fixes R to $\langle R_1,...,R_n \rangle$. This raises the following question. What condition must a function satisfy for being a collective choice rule (in the context of relationals)? Or, in other words, what condition must a function f satisfy so that R= fo $\langle R_1,...,R_n \rangle$ and f: $\mathbb{PE}_{\langle R_1,...,R_n \rangle} \to \mathbb{PE}_R$ are sufficient for R being wholly dependent on $\langle R_1,...,R_n \rangle$? In chapter 4 we introduced two notions, viz. isomorphism preservation and relative effectiveness, and we shall now see whether any of them is pertinent in this context.

5.3 Isomorphism preservation, subordination and social choice

We begin this section with some remarks about the relation between isomorphism preservation of collective choice rules and subordination of the group preferences to the individual preferences.

If R is subordinate to $\langle R_1,...,R_n \rangle$ then the transition $\langle R_1,...,R_n \rangle^R$ from $\langle R_1,...,R_n \rangle$ to R is a function on $\mathbb{C}_{\langle R_1,...,R_n \rangle}$ onto $\mathbb{C}_R$. $\langle R_1,...,R_n \rangle^R$, or more precisely, the restriction g of $\langle R_1,...,R_n \rangle^R$ to $\Pi^{(n)}$ is a function on $\Pi^{(n)}$ into Ω. It is obvious that g fixes R to $\langle R_1,...,R_n \rangle$ and that g is isomorphism-preserving, i.e. if φ is an isomorphism from $\langle A,\rho_i \rangle$ to $\langle B,\sigma_i \rangle$ for all i, $1 \leq i \leq n$, then φ is an isomorphism from $g(\langle A,\rho_1,...,\rho_n \rangle)$ to $g(\langle B,\sigma_1,...,\sigma_n \rangle)$.

If R is fixed to $\langle R_1,...,R_n \rangle$ by a function f, then it is not necessarily the case that R is subordinate to $\langle R_1,...,R_n \rangle$—R could easily violate the isomorphism preservation condition. But if f is isomorphism-preserving relative to $R_1,...,R_n$, i.e. if

$$\varphi \in \mathbb{I}(\langle A,\rho_i \rangle,\langle B,\sigma_i \rangle) \text{ for all } i, \ 1 \leq i \leq k,$$

implies

$$\varphi \in \mathbb{I}(f(A,\rho_1,...,\rho_n), f(B,\sigma_1,...,\sigma_n)),$$

then R (fixed by f) is subordinate to $\langle R_1,...,R_n \rangle$. That is easy to see because $\{\langle A,R_1(A),...,R_n(A) \rangle \,|\, A \in \mathbb{D}\}$ is a subset of the domain of f.

Is the converse also true? Is it the case that if f is not isomorphism-preserving, then R is not subordinate to $\langle R_1,...,R_n \rangle$? In a straight forward way the answer is no. If f is not isomorphism-preserving, then there are $\mathbf{A},\mathbf{B} \in \Pi^{(n)}$ and $\varphi \in \mathbb{I}(\mathbf{A},\mathbf{B})$ such that $\varphi \notin \mathbb{I}(f(\mathbf{A}),f(\mathbf{B}))$. But it could be the case that at least one of $\mathbf{A}$ and $\mathbf{B}$ belongs to $\mathbb{PE}_{\langle R_1,...,R_n \rangle} \setminus \mathbb{E}_{\langle R_1,...,R_n \rangle}$, i.e. at least one of $\langle R_1,...,R_n \rangle(\mathbf{A}) \neq \mathbf{A}$ and $\langle R_1,...,R_n \rangle(\mathbf{B}) \neq \mathbf{B}$ holds. Thus R can be isomorphism-preserving relative to $\langle R_1,...,R_n \rangle$ even if f is not isomorphism-preserving, because the structures that are contrary to the isomorphism preservation of f are not relevant for the isomorphism preservation of R. This is the sense in which the question at the beginning of this paragraph must be answered negatively. But there still remains something to be said.

Whether the relational R fixed by f is subordinate to $\langle R_1,...,R_n \rangle$ or not depends, as we have seen, on which values $\langle R_1,...,R_n \rangle$ takes for different arguments. But the idea behind social choice theory is to study properties of collective choice rules and the properties they induce on the social preference relational. Here the formal properties of the individual preference relationals are relevant but the exact values of the individual preference relationals for different arguments are irrelevant. We will of course be able to discuss the properties of the social preference relational without knowing the extensions of the individual preference relationals. The construction of the collective choice rules are such that they satisfy these requirements: they are defined for the possible extensions of $\langle R_1,...,R_n \rangle$, and $f(A,\rho_1,...,\rho_n)$ is the extension on A of R fixed by f if $\langle A,\rho_1,...,\rho_n \rangle$ is the extension of $\langle R_1,...,R_n \rangle$ on A. If f is not isomorphism-preserving then there is $A,B \in \Pi^{(n)}$ and $\varphi \in \mathbb{I}(A,B)$ such that $\varphi \notin \mathbb{I}(f(A),f(B))$. If $\langle R_1,...,R_n \rangle$ were such that it would take A as the extension on A, and simultaneously, B as the extension on B (which is consistent with the formal properties of $\langle R_1,...,R_n \rangle$), then R would not be subordinate to $\langle R_1,...,R_n \rangle$. Therefore, to ensure that R fixed by f is subordinate to $\langle R_1,...,R_n \rangle$, it seems in a certain sense reasonable to demand f to be isomorphism-preserving.

This line of thought can also be expressed in the following way. Let f^o be the isomorphic closure of f, that is

$$f^o = \{\langle Y,Z \rangle \in \mathbb{C}_{\langle R_1,...,R_n \rangle} \times \mathbb{C}_R \mid Y=Z \; \& \; \exists \langle A,B \rangle \in f: \exists \varphi \in \mathbb{B}i(Y,A): \varphi \in \mathbb{I}(Y,A) \; \& $$
$$\varphi \in \mathbb{I}(Z,B)\} .$$

f^o is obviously a correspondence on $\mathbb{C}_{\langle R_1,...,R_n \rangle}$ to $\mathbb{C}_R$. If f fixes R to $\langle R_1,...,R_n \rangle$ then the transition from $\langle R_1,...,R_n \rangle$ to R is a subset of f^o. Suppose now that the relational S_i has the same range of definition and the same characteristic class as R_i and that the same holds for S and R. If f fixes S to $\langle S_1,...,S_n \rangle$ then the transition from $\langle S_1,...,S_n \rangle$ to S is a subset of f^o. That implies that if f fixes the social preference relational to the individual preference relationals, then the transition from the individuals' to the social preference relational is a subset of the isomorphic closure of f, and that holds whichever the individuals' preference relationals are insofar as their domain is $\mathbb{D}$ and their characteristic class is $\mathbb{C}_{R_i}$. f^o is isomorphism-preserving iff f is and iff every subset of f^o is. If f is isomorphism-preserving then we can be sure that the transition from the individuals' preference relationals (whichever they are) to the social preference relational fixed by f is isomorphism-preserving, and thus the social preference relational fixed by f is subordinate to the individuals' preference relationals.

Let us now turn to the question whether it is reasonable to demand that the group preference relation is subordinate to the individual preference orderings, and if it is therefore also reasonable to demand that the collective choice rules are isomorphism-preserving. The answer seems to be no. The reason is simply that there are functions which are not isomorphism-preserving but still ought to be called collective choice rules. Let us look at an example.

Suppose that f_1 and f_2 both are isomorphism-preserving collective choice rules. Define f in the following way:

$$f(A_1,...,A_n) = \begin{array}{l} f_1(A_1,...,A_n) \ \text{ if } \ A \in \mathbb{D}_1 \\ \\ f_2(A_1,...,A_n) \ \text{ if } \ A \in \mathbb{D}_2 \end{array}$$

where $\langle \mathbb{D}_1, \mathbb{D}_2 \rangle$ is a partition of $\mathbb{D}$. It could very well be the case that f is not isomorphism-preserving. Collective choice rules used in for example parliaments are often built up in this way from isomorphism-preserving choice rules, for instance simple majority rule and some kind of qualified majority rule. Whether simple or qualified majority rule is used then depends on the character of the alternatives, for example constitutional amendments are in some parliaments decided with a qualified majority.

Isomorphism-preserving collective choice rules utilizes only information about the structure of the extensions of individual preference relations. A collective choice rule which is not isomorphism-preserving uses information of something else, too. Given such a collective choice rule it is natural to ask what this information is and what it is about. The information f in the example above utilizes is, besides information of the structure of the argument, also information about which subset of the partition of $\mathbb{D}$ the base set of the argument belongs to.

Another kind of information non-isomorphism-preserving collective choice rules may utilize is about interpersonal comparison of utilities. Even if for all i, $1 \leq i \leq n$, φ is an isomorphism from $R_i(A)$ to $R_i(B)$, $\varphi[f(R_1(A),...,R_n(A))]$ could still differ from $f(R_1(B),...,R_n(B))$ because the interpersonal comparison concerning A may be different from the interpersonal comparison concerning B.

The function f introduced above, which is built up from two isomorphism-preserving collective choice rules f_1 and f_2, need not be isomorphism-preserving. However, it is automorphism-preserving, which is easy to see. Automorphism preservation is therefore perhaps a more interesting condition on collective choice rules than isomorphism preservation.

The condition of isomorphism preservation (or perhaps better automorphism preservation) is not overlooked in the theory of social choice. Something at least related to it has been studied for example by Hansson under the name of neutrality between alternatives (see Hansson, 1969, p. 51) and by Sen under the name of neutrality (see Sen,1970, p.p. 68, 72). If we take Hansson's formulation of the condition and adjust it to the framework adopted here when the preference profiles have X as domain, it can be expressed as follows. A collective choice rule f satisfies *neutrality between alternatives* if the alternatives x and y change places everywhere in a preference profile, they will change places in the corresponding social preference ordering, too (i.e. in the value of f for that profile). We denote this condition NA. Note that NA is a condition which collective choice rules may or may not satisfy. It is usually not treated as a condition which a function must satisfy for being a collective choice rule.

To see the simlarity between NA and isomorphism preservation we reformulate NA somewhat. Suppose $\rho \in \Pi^{(n)}$ and let ρ' be obtained by changing the places of x and y everywhere in ρ. Then $\rho' \in \Pi^{(n)}$ and the function $\varphi_{x,y}: X \rightarrow X$ such that

$$\varphi_{x,y}(z) = \begin{array}{ll} y & \text{if } z=x \\ x & \text{if } z=y \\ z & \text{if } z \neq x,y \end{array}$$

is an isomorphism from ρ to ρ'. ρ' can appropriately be written $\varphi_{x,y}[\rho]$. NA can thus be restated as follows: f satisfies NA iff for every $\rho \in \Pi^{(n)}$ and every $x,y \in X$, $\varphi_{x,y}$ is an isomorphism from $f(\rho)$ to $f(\varphi_{x,y}[\rho])$. Compare this with isomorphism preservation for f: f is isomorphism-preserving iff for every $\rho,\rho' \in \Pi^{(n)}$ and every isomorphism φ from ρ to ρ' it holds that φ is an isomorphism from $f(\rho)$ to $f(\rho')$. It is obvious that the isomorphism preservation of f implies that f satisfies NA.

The expression "is a function of" is frequently used in social choice theory; see for example the first quotation from Sen and the quotation from Arrow at the beginning of section 5.1. The use of the expression has a long tradition, and it is still often used, perhaps especially by social scientists. But it is nowadays frequently regarded as vague or metaphorical (especially outside the realm of connections between numerical measures). What a function is is well known; what it is for something to be a function of something else is not always taken to be well defined. It depends, I think, to a large extent on the fact that "to be a function of" is a relationship between variables in the old sense of variable quantities or magnitudes. And logic and philosophy today has not much room for this concept of a variable. Karl Menger has however pointed out that variables of this kind should be regarded as assignments, maps or functions. In for example the expression "stochastic variable" the old use of "variable" still survives; a stochastic variable is of course a function. (For Menger's view see for instance Menger, 1954, 1956 and 1961.) We shall not discuss here Menger's interesting idea but just point out that since relationals are functions, they can in a Mengerian way be regarded as variables (in the sense of variable quantities). What should it mean, then, that a relational Q is a function of a relational P? Bearing in mind that "to be a function of" expresses complete determination, there seems to be a natural answer: Q is subordinate to P. Let us look at this.

Suppose Q is subordinate to P. What holds for the elements in A with respect to P then completely determines what holds for the elements in A with respect to Q. It is the structure of the extension of P on A which determines [the structure of] the extension of Q on A. (This was discussed in some detail in section 4.8.) If Q is subordinate to P then—and only then—it holds that P^Q is a function. Note that if P^Q is a function then $Q=P^Q \circ P$, where $\circ$ is functional composition.

To interpret the expression "is a function of" as subordination is, in a sense, to understand the expression in a structural way, and I think this is one natural interpretation. But note that, as we saw above, in many contexts it is not reasonable to demand that the group preference relation is a function of the individual preference orderings, when "function of" is interpreted structurally. One may therefore wonder if there is another way of interpreting "to be a function of" and we shall look at this.

5.4   Relative effectiveness, derivability and social choice

Suppose R is fixed to $\langle R_1,...,R_n \rangle$ by f. Then f is a function on $\mathbb{PE}_{\langle R_1,...,R_n \rangle}$ into $\mathbb{PE}_R$ such that $R = f \circ \langle R_1,...,R_n \rangle$. This is a necessary condition for f being a derivation of R from $\langle R_1,...,R_n \rangle$. But it is not sufficient. To be a derivation f must also be relatively effective. Is this an adequate demand of a collective choice rule? At least some of the ideas social theorists have expressed indicate that they have something like it in mind. The very term "collective choice *rule*" may be taken as positive evidence. Furthermore, a collective choice rule is often interpreted as a procedure for passing from the individual preferences to the social preference. (See Luce & Raiffa, 1957, p. 331.) And this procedure seems to be thought of as in a sense formal, i.e. as not having an empirical character. It is then a function not just in the sense of a univalent correspondence; it also has a mathematical or formal character in the sense that it does not involve empirical operations. Rather, it transforms the individual preferences into the social preference. It seems to me that this idea of a collective choice rule as a formal or "non-empirical" rule is fairly common in social choice theory.

There is another idea of collective choice rule supplementing that of "non-empiricalness". One frequently expects it to be possible to [actually] determine the values of a collective choice rule f for given arguments, i.e. if $R = fo \langle R_1,...,R_n \rangle$ then the extension of R on A can be determined by the extension of $\langle R_1,...,R_n \rangle$ on A. If $\mathbb{U}$, i.e. $\cup \mathbb{D}$, is finite, it is quite clear what is meant by this. When $\mathbb{U}$ is infinite, the idea of determinateness must be refined somewhat. The notion of relative effectiveness presented in section 4.6 is an attempt to capture this ideas of "non-empiricalness" and determinateness in the context of relationals and extend them to the infinite case.

If we accept that R is fixed to $\langle R_1,...,R_n \rangle$ by f and f is relatively effective, then f is a derivation of R from $\langle R_1,...,R_n \rangle$ and R is derivable from $\langle R_1,...,R_n \rangle$. Is R also derived from $\langle R_1,...,R_n \rangle$, i.e. is R defined as $fo \langle R_1,...,R_n \rangle$? This is not necessarily the case. R may be derivable from $\langle R_1,...,R_n \rangle$ by another function g, too, and R may be defined as $go \langle R_1,...,R_n \rangle$. Then g and f coincide on $\mathbb{E}_{\langle R_1,...,R_n \rangle}$, i.e.

$$g(A) = f(A)$$

for all A in $\mathbb{E}_{\langle R_1,...,R_n \rangle}$, but outside $\mathbb{E}_{\langle R_1,...,R_n \rangle}$ they may differ. If they coincide on the whole $\mathbb{PE}_{\langle R_1,...,R_n \rangle}$ they are extensionally equivalent but may still be two different functions-in-intension, which may not be unimportant when we use them in the definition of relationals.

Even if the fact that R is derivable from $\langle R_1,...,R_n \rangle$ with f as the derivation does not imply that R is derived from $\langle R_1,...,R_n \rangle$ by f, it may still be the case that R is derived from $\langle R_1,...,R_n \rangle$, but then by another derivation. It is therefore natural to ask if a social preference relational always is derived from the individuals' preference relationals. We have already touched upon this question at the end of section 5.2. I think it is at least not alien to social choice theorist that the social preference is by definition the result of the application of a collective choice rule to the individual preferences. But the theory of social choice seems to be rather heterogeneous, and I do not want to maintain that this is the only existing conception.

Let us now return to the expression "being a function of". It seems to me that it is a fairly vague notion. In the preceding section the structural interpretation of it was discussed. There is perhaps another way to interpret it too, namely as derivability. If Q is derivable from P then there is a relatively effective function f defined for $\mathbb{PE}_R$ such that Q=foP. When we say that Q is a function of P, I think it is possible in some contexts that we thereby mean, that there is a "non-empirical", mathematical transformation of the value of P for an argument to the value of Q for that argument. It may not, therefore, seem unnatural that one explication of "Q is a function of P" is that Q is derivable from P. One can of course object that the condition of relative effectiveness is too strong a condition in this context. The class of mathematical functions are much wider than the class of effective functions (as was already pointed out in connection with derivability in section 4.8). However, it is perhaps also possible that "being a function of" can be understood as a "mixture" of the structural explication and the derivability explication, in analogy with how a derivation in the weaker sense is characterized in section 4.8.

Note that if "being a function of" is understood as derivability, it seems reasonable to demand in many contexts that the social preference relation should be a function of the individual preference orderings.

As has been pointed out several times already, the existence of a function f such that R=fo$\langle R_1,...,R_n \rangle$ does not guarantee that R is in any sense connected to $\langle R_1,...,R_n \rangle$ apart from the fact that the range of definition of $\langle R_1,...,R_n \rangle$ is a subset of the range of definition of R. We are therefore led to consider conditions on such functions. This may not seem in agreement with fundamental ideas held by choice theorists, for example the idea that the dependence between individual and social preferences consists in the existence of a function relating the individual preferences and the social preference. It is perhaps worth emphasizing here that by regarding preference relations as relationals we have changed the framework for the investigation somewhat, and it is in this new framework we have discussed the kind of dependence between social and individual preferences.

5.5 Stability and background for collective choice rules

The most debated condition on a collective choice rule seems to be the condition of independence of irrelevant alternatives. Within the framework of collective choice rules as functions from the set of preference profiles with X as domain this condition can be formulated as follows. The collective choice rule f satisfies the condition of independence of irrelevant alternatives (IIA) if and only if for all preference profiles $\langle \rho_1,...,\rho_n \rangle$, $\langle \rho_1',...,\rho_n' \rangle$ and all subsets A of X: $\rho_i/A = \rho_i'/A$ for all i, $1 \leq i \leq n$, implies

$$f(\rho_1,...,\rho_n)/A = f(\rho_1',...,\rho_n')/A \ .$$

If we let the collective choice rules be functions from the set of preference profiles with elements in $\mathbb{D}$ as domains, then the condition can be given the following wording: The collective choice rule f satisfies (IIA) iff for all preference profiles $\langle A,\rho_1,...,\rho_n \rangle$ and $\langle A',\rho_1',...,\rho_n' \rangle$: $\rho_i/A' = \rho_i'/A$ for all i, $1 \leq i \leq n$, implies

$$f(\langle A,\rho_1,...,\rho_n \rangle)/A' = f(\langle A',\rho_1',...,\rho_n' \rangle)/A \ .$$

(Note that if $A \cap A' = \varnothing$, then $\rho_i/A' = \varnothing$ and $f(\rho_1,...,\rho_n)/A' = \varnothing$, and analogously $\rho_i'/A = \varnothing$ and $f(\rho_1',...,\rho_n')/A = \varnothing$.) If we denote $\langle A,\rho_1,...,\rho_n \rangle$ by A and $\langle A',\rho_1',...,\rho_n' \rangle$ by A', then the condition (IIA) can be expressed as follows: For all A and A' in $\mathbb{PE}_{\langle R_1,...,R_n \rangle}$,

$$A/A' = A'/A \ \Rightarrow \ f(A)/A' = f(A')/A \ .$$

Collective choice functions are obviously s-functions, so the notions of stability and weak stability defined in section 4.7 (and monotonicity defined in section 7.6) can be applied to collective choice rules. It is immediately seen that (IIA), as we have formulated it here, is equivalent to f being stable. There are reasons, I think, that (IIA)—when transformed into the framework of relationals—ought not to be formulated as above but in the following way: For all A and A' in $\mathbb{PE}_{\langle R_1,...,R_n \rangle}$,

$$A/A' = A'/A \ \& \ A/A' \in \mathbb{PE}_{\langle R_1,...,R_n \rangle} \ \Rightarrow \ f(A)/A' = f(A')/A \ .$$

This condition is equivalent to weak stability. If $\mathbb{PE}_{\langle R_1,...,R_n \rangle}$ is closed under formation of substructures, then strong and weak stability coincide. It is not important here exactly which condition be termed "independence of irrelevant alternatives", only that (IIA) is closely related to stability.

The condition of independence of irrelevant alternatives (IIA) is as we have seen closely related to the idea of stability. Within the framework of preferences represented by relationals we are able to distinguish between stability for collective choice rules and for social preference relationals. Suppose $R = fo\langle R_1,...,R_n \rangle$. Then R is stable relative to $\langle R_1,...,R_n \rangle$ if f is stable. It is usually regarded as desirable that the social preference relational is stable relative to the individual preference relationals, and therefore it is desirable that the collective choice rule is stable. But why should R be stable? If R is not stable both

$$R(A;x,y) \ \& \ \neg R(A;y,x)$$

and

$$R(B;y,x) \ \& \ \neg R(B;x,y)$$

may be true propositions. Then we are not able to tell whether x is socially preferred to y or not—it depends on which domain we apply R to. However, we are usually interested in what holds irrespective of the domain we consider. The extension of R over a set of alternatives is often thought of as having strong implications for how society should act. However, it is not always clear exactly what is the relevant set of alternatives. If R holds between x and y in all domains containing x and y we have a desirable situation. In some contexts we may be satisfied with just "half stability or less", i.e. either upward or downward stability, or stability upward or downward for a point from a certain domain (see section 3.3).

In Danielsson (1974)—see p. 31—(IIA) is divided into two conditions, viz. independence of actual alternatives and independence of possible alternatives. To do justice to this distinction when applied to relationals, it is appropriate to regard relationals as binary functions, i.e. as functions which take ordered pairs of sets as arguments. Such binary relationals, which I shall call *two-dimensional,* turn out to have an interest in their own right besides their use for expressing the Danielsson distinction within the frame of relationals. A two-dimensional relational Q takes ordered pairs $\langle X,A \rangle$ of elements X and A in $\mathbb{D}_Q$ such that $A \subseteq X$ as arguments. If A and X are in $\mathbb{D}_Q$ and $A \subseteq X$, then $Q(X,A)$ is a structure with A as base set and is interpreted as the extension of Q on A with X as background. We often write $\langle x_1,...,x_n \rangle \in Q(X,A)$ as $Q(X,A;x_1,...,x_n)$. If $Q(X,A;x_1,...,x_n)$ then Q holds for $x_1,...,x_n$ (in that order) when $x_1,...,x_n$ are regarded as elements in A and with X as background. We now distinguish between domain stability and background stability for Q. Q is *domain-stable* if for all X, A and B in $\mathbb{D}_Q$ such that $A,B \subseteq X$,

$$Q(X,A)/B = Q(X,B)/A .$$

Q is *background-stable* if for all X, Y and A in $\mathbb{D}_Q$ such that $A \subseteq X$ and $A \subseteq Y$,

$$Q(X,A) = Q(Y,A) .$$

It is possible to define weak, upward and downward domain and background stability and, furthermore, study isomorphism preservation for two-dimensional relationals, but we shall not pursue this line of thought any further in this essay. Here we confine ourselves to the following observation. In the account of social choice theory given in section 5.1, the set X played a special role. It is often thought of as the set of conceivable alternatives or the like. When preferences are represented by relationals, it seems natural to consider X, the set of conceivable alternatives in a situation, as a background. In section 5.1 X has the status of a parameter, i.e. an arbitrary constant fixed within a situation but (implicitly) with different interpretations in different situations. To regard X as a background is more like treating X as an (explicit) variable.

5.6 Structural dependence and aggregation; a preliminary remark

A way of looking at social choice theory is as dealing with methods for aggregating or amalgamating individual preference relationals to the social preference relational. Viewed this way, social choice theory is a special case of a general theory of aggregation or amalgamation. Such a theory treats the aggregation or amalgamation of different relationals or systems of relationals—*the factors* (which are not necessarily individual preference relationals)—into a relational or a system of relationals—*the aggregate* (or amalgam). This more general theory has applications to a wide range of phenomena. One important application is in multidimensional decision theory, which deals with decisions when several aspects or dimensions of the alternatives are taken into consideration. Another area where aggregation plays a central role is in measurement theory, for example in the theory of conjoint measurement and in dimensional analysis.

A central question for a study of aggregation is the relationship between the aggregate and the factors, and an important aspect of this question is what kind of dependence exists between them. The aggregate seems not in general to be subordinate to the factors. This might be interpreted as if the aggregation sometimes uses more information than is contained in the structure of the extensions of the factors, and it seems well motivated to ask what kind of information this is. The aggregate is perhaps structurally determined (subordinate) to the factors and this extra information. This question is related to another: To what extent can correlation be analysed in terms of subordination? I.e., if S is correlated with R, under what circumstances can this be explained as subordination of one system S' to another R' where S' is related to S and R' related to R in a special way? For both these questions interfactorial comparison—the generalized counterpart to interpersonal comparison of utility in social choice theory—will probably play a central role.

CHAPTER 6

CONFORMITY AND MEASURES

6.0 Introduction

Homomorphic representations of relational structures into systems of real numbers is often held to be the kernel of measurement theory. Scales and measures are usually regarded as essentially such homomorphic representations. Let us say, as is customary in measurement theory, that the function h on A into B is a *homomorphism* on (or of) the relational structure $\langle A, \rho \rangle$ into $\langle B, \sigma \rangle$ iff for all $x_1, \ldots, x_n \in A$,

$$\rho(x_1, \ldots, x_n) \text{ iff } \sigma(h(x_1), \ldots, h(x_n)) .$$

A homomorphic representation of a relational into a structure is a more complicated notion, and a measure or scale for a relational will be a homomorphic representation of a special kind.

In section 6.3 we study how the notion of homomorphism for relations can be generalized to relationals. To avoid confusion I intend to use "homomorphic representation" in connection with relationals and "homomorphism" in connection with structures. In section 6.4 the concept of a measure is adjusted to make it applicable to relationals. Some of the notions already used in the study of relationals, for example stability, will also be applicable to homomorphic representations and measures; this is considered in sections 6.3 and 6.4. Section 6.3 is devoted to the special case of numerical representations and measures. The main interest in representations and measures is—within the context of this study—how representations and measures can be used to express connections between relationals. The fundamental instruments for this investigation are the ideas of equality preservation, where equality is understood with regard to an aspect, and of the independent realizability of aspects. These ideas are introduced in section 6.1 within a standard measurement theoretical framework. And in section 6.2 we apply them to relationals. The study of dependence expressed by measures and representations are taken up in section 6.6.

6.1. Equality preservation and independent realizability

In this section we take a first look at the idea of equality preservation and independent realizability within a traditional measurement theoretical framework. In a later section the idea is generalized to the context of relationals.

Let us return to the example from physics in the beginning of section 1.1. Suppose v is a vessel filled with liquid. Let

F be the force on the bottom of v in N (Newtons)

h be the depth of the liquid of v in m (meters)

A be the bottom area of v in m^2

d be the density of the liquid of v in kg/m^3

g the acceleration at free fall in m/s^2.

At every place on the surface of the earth the value of g is, to one decimal place, $9.8\,m/s^2$; g is taken as a constant in the discussion of this example.

It holds that

$$F = A \cdot h \cdot d \cdot g.$$

This is an example of a numerical law; more exactly a physical law. It is formulated for an arbitrary vessel v, which implies that it holds for the class V of all liquid filled vessels.

Note that

the force on the bottom of ... in N

is a mapping which assigns a number to every vessel v that is the force acting on its bottom in Newtons. Let us denote this assignment by $\underline{F}$. $\underline{F}$ is thus a function on V into the set Re+ of positive real numbers. By analogy, let

$\underline{h}$ be the assignment *the depth of the liquid of ... in m*

$\underline{A}$ be the assignment *the bottom area of ... in m²*

$\underline{d}$ be the assignment *the density of the liquid of ... in kg/m³*.

$\underline{h}$, $\underline{A}$ and $\underline{d}$ are thus also functions on V into Re+. We can write the law

For all v in V: $\underline{F}(v) = \underline{A}(v) \cdot \underline{h}(v) \cdot \underline{d}(v) \cdot g.$

Let ϕ be the function on $(Re^+)^3$ into Re+ such that

$$\phi(x,y,z) = x \cdot y \cdot z \cdot g.$$

We can now write the law

For all v in V: $\underline{F}(v) = \phi(\underline{A}(v), \underline{h}(v), \underline{d}(v)).$ (6.1.1)

By using what is sometimes called the complex mapping (see Kuratowski, 1972, p. 63) and sometimes the restricted [Cartesian] product (Hu, 1964, p. 13), it is possible to get rid of the arguments in 6.1.1. The *restricted product* of the functions f and g with the same domain D is the function h defined on D such that

$$h(x) = \langle f(x), g(x) \rangle.$$

Let us denote h by $\langle f, g \rangle$. Thus

$$\langle f, g \rangle(x) = \langle f(x), g(x) \rangle$$

and $\langle f, g \rangle$ is the restricted product of f and g. 6.1.1 can now be written

$$\underline{F} = \phi \circ \langle \underline{A}, \underline{h}, \underline{d} \rangle. \tag{6.1.2}$$

In 6.1.2 the numerical law is expressed as an equality between functions, or to be more exact, as an equation with functions as terms. ϕ has 3-tuples of positive real numbers as arguments and $\underline{F}$, $\underline{A}$, $\underline{h}$, and $\underline{d}$ are functions with vessels as arguments and positive real numbers as values. ϕ is thus the only function in the law which has mathematical objects both as arguments and values.

$\underline{F}$, $\underline{h}$, $\underline{A}$ and $\underline{d}$ are measures of, respectively, the force on the bottom (**F**), the depth of the liquid (**H**), the bottom area (**A**) and the density of the liquid (**D**). According to a traditional view in measurement theory the aspect "force on the bottom", i.e. **F**, is a set theoretical structure with V as base set. The measure $\underline{F}$ is a homomorphism on this structure into a structure with the positive real numbers as domain. The situation is similar with **H** and $\underline{h}$, **A** and $\underline{A}$ and **D** and $\underline{d}$. The numerical law (6.1.2) expresses a kind of connection between the aspects **F**, **H**, **A** and **D** formulated in terms of measures.

According to the tradition in measurement theory referred to above, a numerical law is in general of the following form:

$$h_0 = f \circ \langle h_1, \ldots, h_n \rangle \tag{6.1.3}$$

where h_i is a measure of the aspect α_i and thus a homomorphism on α_i into a numerical structure N_i. Note that all α_i have the same domain, for example A. (I disregard here the fact that numerical laws can also have the more complex form $g \circ \langle h_{n+1}, \ldots, h_{n+m} \rangle = f \circ \langle h_1, \ldots, h_n \rangle$.)

It seems reasonable to say that a numerical law expresses a functional dependence between measures. 6.1.2 implies that $\underline{F}$ is a function of $\underline{A}$, $\underline{h}$ and $\underline{d}$. And 6.1.3 says that h_0 is a function of $h_1, \ldots h_n$. Let us now turn to the question what kind of dependence between aspects a functional dependence between measures implies.

Suppose that for all i, $0 \leq i \leq n$, α_i has A as its domain. Let $\sim_i$ denote equality with respect to α_i. $a\sim_i b$ thus means that a and b are equal with respect to α_i. We now suppose that for all i, $0\leq i\leq n$,

$$a\sim_i b \text{ iff } h_i(a)=h_i(b).$$

Then there is a function f such that $h_0 = f\circ\langle h_1,...,h_n\rangle$ iff for all $a,b\in A$,

$$a\sim_i b \text{ for all i, } 1\leq i\leq n, \text{ implies } a\sim_0 b. \qquad (6.1.4)$$

This is easy to see. Suppose first that $h_0 = f\circ\langle h_1,...,h_n\rangle$ and that $a\sim_i b$ for all i, $1\leq i\leq n$. Then $h_i(a)=h_i(b)$ for all i, $1\leq i\leq n$. From this follows that $f(h_1(a),...,h_n(a)) = f(h_1(b),...,h_n(b))$ and thus $h_0(a)=h_0(b)$, which implies that $a\sim_0 b$. Suppose now that 6.1.4 holds for all $a,b\in A$. Then

$$h_i(a)=h_i(b) \text{ for all i, } 1\leq i\leq n, \text{ implies } h_0(a)=h_0(b)$$

and thus $f=\{\langle\langle h_1(a),...,h_n(a)\rangle,h_0(a)\rangle \mid a\in A\}$ is a function. It then follows immediately that $h_0 = f\circ\langle h_1,...,h_n\rangle$.

That 6.1.4 holds for all $a,b\in A$ expresses the idea of *equality preservation*; simultaneous equality with respect to all α_i, $1\leq i\leq n$, is preserved in equality with respect to α_0. Let us say that if 6.1.4 holds for all $a,b\in A$, then α_0 is *equality-preserving* relative to $\alpha_1,...,\alpha_n$. As we have just seen, equality preservation holds iff there is a functional dependence between the corresponding measures.

Let α and β be aspects, as above characterized by relational structures, and with the same base set A. We say that a with regard to α and b with regard to β are *simultaneously realizable* if there is $c\in A$ such that $a\sim_\alpha c$ and $b\sim_\beta c$. It is easy to see that a with regard to α and b with regard to β are simultaneously realizable iff a and b are related by the relative product of $\sim_\alpha$ and $\sim_\beta$, i.e. $a(\sim_\alpha \mid \sim_\beta)b$.

We further say that two aspects α and β are *independently realizable* if for all $a,b\in A$, a with regard to α and b with regard to β are simultaneously realizable. Thus α and β are independently realizable iff $(\sim_\alpha \mid \sim_\beta)=A\times A$. This notion is based on the idea of independent realizability found in Krantz et. al. (1971) p. 246 and reproduced in section 1.9 of this book.

The law 6.1.1 expresses a dependence between the force on the bottom on the one hand and the area of the bottom, the height and the density of the liquid on the other: the force preserves equality of the area, the depth and the density. But note that given a specific area and depth we can get whatever force we like by choosing a suitable density: If the force is F Newtons, the area is A m^2 and the depth is h kg/m^3 then choose the density as F/Ahg. Therefore, given two vessels v and w, we can (in principle) find a vessel x such that $v\sim_F x$ and $w\sim_{AH} x$, where $w\sim_{AH} x$ iff $w\sim_A x$ and $w\sim_H x$. This amounts to saying that the force on the one hand and the area and depth on the other are independently realizable.

Analogously, the force is independently realizable with respect to area and density (taken together) and likewise with respect to depth and density taken together. It is also the case that the force is realizable independently of area, depth and density each taken separately.

It seems to hold quite generally that if $h_0 = f\!\propto\!\langle h_1,...,h_n\rangle$ is an elementary physical law, where h_i is a measure of α_i, then α_0 preserves equality with respect to $\alpha_1,...,\alpha_n$ (i.e. $x\!\sim_0\!y$ if $x\!\sim_i\!y$ for all i, $1\leq i\leq n$). It also seems to hold that for all i,j such that $0\leq i,j\leq n$ and $i\neq j$, α_i and α_j are independently realizable. And furthermore, α_0 is independently realizable with respect to every system of aspects consisting of a proper subset of $\{\alpha_1,...,\alpha_n\}$.

Equality preservation is as we have seen equivalent to functional dependence between measures but expresses also a kind of dependence between aspects. Independent realizability is of course a condition of independence between aspects We can in fact see equality preservation and independent realizability as endpoints on a dependence scale. There is of course a difference between this scale and the one with subordination and uncorrelation as endpoints, since equality preservation and independent realizability—as they are treated in this section—hold for structures while subordination and uncorrelation hold for systems of relationals. But even if we disregard this difference for the moment, the two scales are of quite different character. Equality preservation and independent realizability are based on what holds for *objects* while subordination and uncorrelation are based on what holds for *structures*. Nevertheless, they are analogous. Although the relative product of $\sim_\alpha$ and $\sim_\beta$ is something quite different from the transition from α to β, the former plays, relative to equality preservation and independent realizability, a role analogous to that which the transition plays relative to subordination and uncorrelation. Subordination can in fact be regarded as preservation of equality between structures. And that **A** and **B** are related by the transition from α to β means in a sense that **A** as an extension of α and **B** as an extension of β are simultaneously realizable. Uncorrelation is therefore a condition of independent realizability based on simultaneous realizability of structures. This is considered in chapter 8.

In this section aspects were characterized as relational structures and within this framework numerical laws, equality preservation and independent realizability were discussed. These notions have their counterparts for systems of relationals, but the situation is somewhat more complicated there and will be considered in the following sections of this chapter.

In section 4.9 the notion of relevance was discussed. Determination was understood there as subordination and various notions of significance were introduced. We can proceed in analogous fashion with determination as equality preservation. Let us for example say that β is *strongly relevant* for γ with respect to α if γ is equality-preserving relative to α and β but γ and α are independently realizable. Let us further say that β is *relevant* for γ with respect to α if $(\sim_\gamma | \sim_{\alpha\beta}) \subset (\sim_\gamma | \sim_\alpha)$, where $x \sim_{\alpha\beta} y$ means $x \sim_\alpha y$ and $x \sim_\beta y$.

In section 1.1 the following two ideas of relevance were introduced:
(1) α is relevant to β if β is determined by α or there are $\alpha_1,...,\alpha_n$ such that β is determined by $\alpha,\alpha_1,...,\alpha_n$ but β is not determined by $\alpha_1,...,\alpha_n$.
(2) α is relevant to β when β is not constant and is determined by α, when everything else is kept constant.

Let us look at how those ideas of relevance can be analysed within the framework of this section, where preservation of equality is taken as the explication of the idea of determination, i.e. α_0 is determined by $\alpha_1,...,\alpha_n$ if α_0 is equality preserving relative to $\alpha_1,...,\alpha_n$. We start off with (1) and say more specifically that α is relevant$_1$ to β with respect to $\alpha_1,...,\alpha_n$ if β is equality preserving relative to $\alpha,\alpha_1,...,\alpha_n$ but β is not equality preserving relative to $\alpha_1,...,\alpha_n$. Note that if β is equality preserving relative to $\alpha,\alpha_1,...,\alpha_n$ but not equality preserving relative to $\alpha_1,...,\alpha_n$ then there are a and b such that $a \approx_i b$ for all i, $1 \leq i \leq n$, but not $a \approx_\beta b$ and not $a \approx_\alpha b$. And conversely, if there are a and b such that $a \approx_i b$ for all i, $1 \leq i \leq n$, but not $a \approx_\beta b$ then β is not equality preserving relative to $\alpha_1,...,\alpha_n$.

We now consider (2). Let us say that α is relevant$_2$ to β with respect to $\alpha_1,...,\alpha_n$ if β is not constant and is equality preserving relative to α, given that $\alpha_1,...,\alpha_n$ are kept constant. That β is not constant given that $\alpha_1,...,\alpha_n$ are kept constant means that there are a and b such that not $a \approx_\beta b$ but $a \approx_i b$ for all i, $1 \leq i \leq n$. That β is equality preserving relative to α given that $\alpha_1,...,\alpha_n$ are kept constant holds if and only if β is equality preserving relative to $\alpha,\alpha_1,...,\alpha_n$. It is therefore easy to see that α is relevant$_1$ to β iff α is relevant$_2$ to β.

6.2 Congruence relationals, conformity and import

In this section the notions of equality preservation and independent realizability discussed in section 6.1 are generalized to relationals. In doing this we need a generalization of the notion of a congruence relation pertaining to relations and structures so that it will hold for relationals, and we start off by considering this.

A relational R with $\mathbb{D}$ as range of definition is an *equivalence relational* on $\mathbb{D}$ iff for all $A \in \mathbb{D}$, R(A) is an equivalence relation on A. Let R and S be equivalence relationals on $\mathbb{D}$. We say that R is *coarser* than S if for all $A \in \mathbb{D}$, R(A) is coarser than S(A), i.e. $R(A) \supseteq S(A)$. If R is coarser then S then S is *finer* than R.

Let R be a relational system with $\mathbb{D}$ as range of definition. A relational κ with $\mathbb{D}$ as range of definition is a *congruence relational* for R if for all $A \in \mathbb{D}$, $\kappa(A)$ is a congruence relation for R(A). From the definition it follows immediately that a congruence relational is always an equivalence relational. The set of congruence relationals for R is denoted $\mathbb{K}_R$. In section 2.10 the notion of a coarsest congruence relation $\sim_A$ for a structure A was introduced. Let $\sim_R$ be the relational with $\mathbb{D}$ as range of definition such that for all $A \in \mathbb{D}$, $\sim_R(A)$ is the coarsest congruence relation for R(A), i.e. $\sim_R(A) = \sim_{R(A)}$. It is easy to see that $\sim_R$ is the coarsest congruence relational for R.

Note that the notions we have introduced above are defined "pointwise" (domainwise) in terms of the corresponding notion for relations and structures.

Let us say that S is *congruence-preserving* relative to R if $\mathbb{D}_R \subseteq \mathbb{D}_S$ and for all $A \in \mathbb{D}_R$,
$$\mathbb{K}(R(A)) \subseteq \mathbb{K}(S(A)).$$
Compare this with the definition of automorphism preservation: S is automorphism-preserving relative to R if $\mathbb{D}_R \subseteq \mathbb{D}_S$ and for all $A \in \mathbb{D}_R$,
$$\mathbb{I}(R(A)) \subseteq \mathbb{I}(S(A)).$$
Congruence preservation can be applied to s-functions in general. Suppose f is an s-function on K into L. Then f is congruence-preserving if for all $X \in K$,
$\mathbb{K}(X) \subseteq \mathbb{K}(f(X))$.

If $\mathbb{D}_R = \mathbb{D}_S$ and S is congruence-preserving relative to R then we say that S *conforms to* R or that S *is in conformity with* R. That S conforms to R is abbreviated as $S conf R$ and the relation which then holds between S and R is called *conformity*. Let us further say that R and S are *equiform* if $S conf R$ and $R conf S$.

In section 2.10 it was shown that
$$\mathbb{K}(B) \supseteq \mathbb{K}(A) \text{ iff } \sim_B \supseteq \sim_A.$$
(See 2.10.5.) This implies that

$$\mathbb{K}(S(A)) \supseteq \mathbb{K}(R(A)) \text{ iff } \sim_S(A) \supseteq \sim_R(A).$$

Suppose that $\mathbb{D}_R = \mathbb{D}_S = \mathbb{D}$. Then $\mathbb{K}(S(A)) \supseteq \mathbb{K}(R(A))$ holds for all $A \in \mathbb{D}$ iff $\sim_S$ is coarser than $\sim_R$. From this it follows that $S conf R$ iff $\sim_S$ is coarser than $\sim_R$. R and S are thus equiform iff $\sim_R = \sim_S$.

The definitions of congruence preservation and conformity above are formulated for relationals, but it is a straightforward matter to generalize the definitions so that they hold of systems of relationals.

One might wonder how *conf* is related to *sub*. It might seem likely that $S conf R$ is closely related to $\sim_S sub \sim_R$. But these two notions do not coincide nor does either imply the other, which the following simple examples show.

Let $\mathbb{D}$ consist of only the set $A = \{a,b,c\}$. Let $R(A) = \{\langle a,a \rangle, \langle b,b \rangle, \langle c,c \rangle\}$ and $S(A) = \{\langle a,b \rangle, \langle b,a \rangle, \langle a,a \rangle, \langle b,b \rangle, \langle c,c \rangle\}$. Then, according to 2.10.11, $R(A)$ and $S(A)$ are equivalence relations so $R(A)$ is the coarsest congruence relation for $R(A)$ and $S(A)$ is the coarsest congruence relation for $S(A)$, i.e. $\sim_R = R$ and $\sim_S = S$. It is therefore obvious that $\sim_S$ is coarser than $\sim_R$ and thus, $S conf R$. Note that $\mathbb{I}(R(A)) = \mathbb{B}i(A)$ but $\mathbb{I}(S(A)) \neq \mathbb{B}i(A)$ since for example $\{\langle a,a \rangle, \langle b,c \rangle, \langle c,b \rangle\} \notin \mathbb{I}(S(A))$ which can be seen from the fact that $a \sim_S b;A$ but $\neg(a \sim_S c;A)$. Thus, it is not the case that $\mathbb{I}(S(A)) \supseteq \mathbb{I}(R(A))$, and therefore not the case that $\mathbb{I}(\sim_S(A)) \supseteq \mathbb{I}(\sim_R(A))$. Note also that $R sub S$ and $\sim_R sub \sim_S$ but $\neg R conf S$. This example therefore shows that $S conf R$ does not generally imply that $\sim_S sub \sim_R$ and that $\sim_S$ is not automorphism-preserving relative to $\sim_R$. It also shows that neither $R sub S$ nor $\sim_R sub \sim_S$ implies $R conf S$.

The above example is a special case of a more general result which illustrates the difference between *sub* and *conf*. Let $\Im$ be the identity relational, i.e. $\Im(A) = \{\langle a,a \rangle \mid a \in A\}$. Note that $\Im sub R$ and $R conf \Im$ for every relational R. But it does not hold for every relational R that $R sub \Im$ and $\Im conf R$.

Congruence preservation is an explication of what in chapter 1 was called equality preservation. There it was maintained that equality preservation is one way of interpreting the idea of determination. If this is correct, *conformity* should express some kind of dependence relation between systems of relationals (dependence in the sense of determination). It can be shown under a fairly general condition that if $S conf R$ then $S corr R$. But in chapter 1 something more was suggested, namely that equality preservation explicates one idea of *complete determination*. Equality preservation was conjectured to mean complete determination on the object level unlike *sub*, which means complete determination on the structure level. Therefore there ought to be a notion of complete undetermination corresponding to *uncorr* but on the object level. In section 6.1 the notion of independent realizability of aspects was explained which seems to be what we are looking for. But how this idea is to be transformed into a well-defined notion applicable to relationals seems problematic. Let us consider this in some detail. We say that a with regard to R and b with regard to S are *simultaneously realizable in* A if there is $c \in A$ such that $a \sim_R c;A$ and $b \sim_S c;A$. Consider the following condition.

(6.2.1)   For all $A \in D$ and all $a,b \in A$, a with regard to R and b with regard to S are simultaneously realizable in A.

It is reasonable to say that this condition is sufficient for R and S being independently realizable. But it seems too strong to be necessary. R and S might be independently realizable although a with regard to R and b with regard to S need not be simultaneously realizable in every domain A to which they belong. It might suffice that for all A in D there is an A+ in D such that $A \subseteq A+$ and for all a and b in A+, a with regard to R and b with regard to S are simultaneously realizable in A+. But then it is important that $R(A+)/A=R(A)$ and $S(A+)/A=S(A)$, so for objects in A it does not matter if we consider them within A or A+. The notion we arrive at in this way we call "unlinked" and it is defined in the following way.

R and S are *unlinked* if for all $A \in D$ the following conditions are satisfied:
(1) there is a domain $A+ \in D$ such that $A \subseteq A+$
(2) for all $a,b \in A+$, a with regard to R and b with regard to S are simultaneously realizable in A+
(3) $R(A+)/A=R(A)$ and $S(A+)/A=S(A)$.

A somewhat different idea of independent realizability is expressed in the following definition of the notion of unconnectedness.

R and S are *unconnected* if for all $A,B \in D$, if $\varphi \in Bi(A,B)$ then
(1) there is $C \in D$ and $\psi \in Bi(A,C)$ such that $a \sim_R \psi(a);A \cup C$ and $\varphi(a) \sim_S \psi(a);B \cup C$, and
(2) $R(A \cup C)/A=R(A)$ and $R(A \cup C)/C=R(C)$, and
(3) $S(A \cup C)/A=S(A)$ and $S(A \cup C)/C=S(C)$.

Let us illustrate the idea behind the definition with the following simple example. Consider the physical aspects length and weight. Suppose A consists of for example three rigid rods a_1, a_2 and a_3 while B consists of three weights b_1, b_2 and b_3. Since length and weight are in an intuitive sense independently realizable we can find (or construct) a set consisting of three objects c_1, c_2 and c_3 (for example rigid rods) such that the length of c_i equals the length of a_i, $1 \leq i \leq 3$, and the weight of c_i equals the weight of b_i, $1 \leq i \leq 3$. Let now $\varphi : A \rightarrow B$ be defined by $\varphi(a_i) = b_i$ and $\psi : A \rightarrow C$ by $\psi(a_i) = c_i$. Then $a_i {\sim}_L \psi(a_i); A \cup C$ and $\varphi(a_i) {\sim}_W \psi(a_i); B \cup C$ where L is the aspect of the length and W the aspect of weight. The idea of length and weight being independently realizable thus finds expression in the existence of the set C satisfying condition (1). (2) and (3) must hold since we make comparison within $A \cup C$ but for objects in A it should not matter if we consider A or $A \cup C$, and the situation is analogous with C.

"Unlinked" and "unconnected" are different formulations of the idea of independent realizability between relationals. I do not want to suggest that they are the only possible explications of this idea, but I will not consider other possibilities here.

If conformity is, as is suggested here, one explication of the idea of complete determination, then we would expect that an explication of relevance in terms of conformity could be given in a way analogous to the way significance was defined in terms of subordination in 3.5. This can be done, and we use the term *import* for this explication of relevance.

By analogy with what was done in 3.6, let us introduce four notions of import. We say that

(1) R has *conf-import* for S with respect to Q if S is in conformity with $\langle Q,R \rangle$ but S is not in conformity with Q

(2) R has *link-import* for S with respect to Q if S and $\langle Q,R \rangle$ are linked but S and Q are unlinked

(3) R has *strong import* for S with respect to Q if S is in conformity with $\langle Q,R \rangle$ but S and Q are unlinked

(4) R has *import* for S with respect to Q if there is $A \in \mathbb{D}$ such that
$$(\sim_{QR}(A)\,|\sim_S(A)) \subset (\sim_Q(A)\,|\sim_S(A)).$$

6.3 Homomorphic representations

Let K be a class of relational structures of the same type τ and let Z be a relational structure similar to the structure in K, i.e. of type τ. A *homomorphic representation* H of K into Z is a function with domain K and with homomorphisms on the argument into Z as values. Thus for A in K, H(A) is a homomorphism on A into Z. Two homomorphic representations H and H' are equal. i.e. H=H', if they are defined for the same class K of structures and H(A)=H'(A) for all A in K. A homomorphic representation of a relational R is a homomorphic representation of $\mathbb{E}_R$.

The notions of stability applied to s-functions have analogues for homomorphic representations. A homomorphic representation H of K into **Z** is *stable* iff for all **A** and **B** in K

$$A/B = B/A \quad \text{implies} \quad H(A)/B = H(B)/A .$$

Further, H is *weakly stable* iff for all **A**,**B** in K such that B⊆A,

$$B=A/B \quad \text{implies} \quad H(B) = H(A)/B .$$

It is easy to see that stability implies weak stability.

Even the idea behind isomorphism preservation has applications to homomorphic representations. Let us say that H is *structure determined* iff for all **A** and **B** in K,

$$\varphi \in \mathbb{I}(A,B) \quad \text{implies} \quad H(A) = H(B) \circ \varphi.$$

The commutative diagram below illustrates the condition.

Let us further say that H' is *iso-preserving* relative to H if H and H' are homomorphic representations of K into **Z** and for all **A** and **B** in K and for all bijections φ from A to B,

$$H(A) = H(B) \circ \varphi \quad \text{implies} \quad H'(A)=H'(B) \circ \varphi.$$

(I have chosen the term "iso-preservation" because the notion resembles isomorphism preservation but is still distinct from it.)

Let us look at an example of a homomorphic representation. Suppose W is the class of all finite weak orders. Define a function β on W such that for all $A=\langle A, \succeq \rangle$ in W and all a in A,

$$\beta(A)(a) = \text{card}\{x \in A \mid a \succ x\} - \text{card}\{x \in A \mid x \succ a\} .$$

$\beta(A)$ is of course a homomorphism on A into $\langle \text{Int}, \geq \rangle$, where Int is the set of integers. Hence, β is a homomorphic representation of W into $\langle \text{Int}, \geq \rangle$. In the tradition from group decision theory (see Gärdenfors, 1973, p. 5) $\beta(A)$ will be called the *Borda representation* of W.

Let us now generalize the notion of Borda representation to k dimensions. A relational structure $\langle X, \rho_1, ..., \rho_k \rangle$ where $\rho_1, ..., \rho_k$ are weak orders on X will here be called a *k-dimensional weak order*. Let $W^{(k)}$ be the class of all finite k-dimensional weak orders. Define, for all $A = \langle A, \succeq_1, .., \succeq_k \rangle$ in $W^{(k)}$, $\beta^{(k)}(A)$ as the restricted product of the functions $\beta(\langle A, \succeq_1 \rangle), ..., \beta(\langle A, \succeq_n \rangle)$, i.e.

$$\beta^{(k)}(A) = \langle \beta(\langle A, \succeq_1 \rangle), ..., \beta(\langle A, \succeq_n \rangle) \rangle$$

where β is the Borda representation. $\beta^{(k)}$ is a homomorphic representation of $W^{(k)}$ into $\langle Int^{(k)}, \geq_1, ..., \geq_k \rangle$ where $Int^{(k)}$ is the set of k-tuples of integers and $\geq_i$ is defined by

$$\langle r_1, ..., r_k \rangle \geq_i \langle s_1, ..., s_k \rangle \text{ iff } r_i \geq s_i .$$

Let us call $\beta^{(k)}$ *the k-dimensional Borda representation* of $W^{(k)}$. β can therefore be called the uni-dimensional Borda representation.

It is easily proved that the k-dimensional Borda representation is structure determined but neither stable nor weakly stable.

6.4 Measures

A measure for the relational R is a function which has the sets in the range of definition $\mathbb{D}$ of R as arguments and functions on the arguments as values. A measure for a relational is always with respect to a relational structure and the value of a measure for the set A in $\mathbb{D}_R$ is a homomorphism on R(A) into the actual relational structure. Thus, if M is a *measure* for R with respect to $Z = \langle Z, \sigma \rangle$, then for all A in $\mathbb{D}_R$,

$$M(A): A \rightarrow Z$$

and for all $a_1, ..., a_n$ in A,

$$R(A; a_1, ..., a_n) \text{ iff } \sigma(M(A)(a_1), ..., M(A)(a_n)) .$$

M(A) can, by analogy with R(A), be called the extension of M on A. M(A)(a) is the value of the measure M for an object a in a domain A. Two measures M and M' for a relational R are equal, i.e. M=M', if M(A)=M'(A) for all A in $\mathbb{D}_R$. We use equality, = , here in such a way that two measures may be equal without being identical. Equality is therefore a kind of extensional equality.

If M is a measure for R with respect to Z, then for all A in $\mathbb{D}_R$,

$$R(A) = M(A)^{-1}[Z] = \langle M(A)^{-1}[Z], M(A)^{-1}[\sigma] \rangle$$

where

$$M(A)^{-1}[Z] = \{y \mid M(A)(y) \in Z\} = A$$

and
$$M(A)^{-1}[\sigma] = \{\langle y_1,...,y_n\rangle \in A^n \mid \sigma(M(A)(y_1),...,M(A)(y_n))\}.$$

The condition of stability for relationals has an analogy for measures. A measure M for R is *stable* iff for all A and B in $\mathbb{D}_R$,
$$M(A)/B = M(B)/A,$$
where $M(A)/B$ is the restriction of the function $M(A)$ to B, i.e. $M(A)/B = \{\langle x,y\rangle \in M(A) \mid x \in B\}$. It is easy to prove the following theorem.

Theorem 6.4.1 If there exist a stable measure for a relational, then the relational is stable.

Note that a stable relational can very well have measures which are not stable.

The condition of stability can be weakened even for measures. A measure M for R is *weakly stable* iff for all A and B in $\mathbb{D}_R$ such that $B \subseteq A$,
$$M(B) = M(A)/B.$$
Stability implies of course weak stability. Theorem 6.4.1 has an analogue for weak stability.

Theorem 6.4.2 If there exists a weakly stable measure for a relational, then the relational is weakly stable.

Note that a weakly stable relational can have measures which are not weakly stable.

The idea of isomorphism preservation can also be applied to measures, but some adjustments must be made by analogy with those for homomorphic representations. A measure M for R is *structure determined* iff for all A and B in $\mathbb{D}_R$,
$$\varphi \in \mathbb{I}(R(A),R(B)) \text{ implies } M(A) = M(B) \circ \varphi.$$
$M(A) = M(B) \circ \varphi$ is equivalent to
$$M(A)(x) = M(B)(\varphi(x))$$
for all x in A . Since $B = \varphi[A]$,
$$M(A)(x) = M(\varphi[A])(\varphi(x)).$$
Note that although structure determination for measures might be desirable in some contexts, it is not generally a desideratum; many natural measures do not satisfy it —often they cannot do so. Suppose M and M' are measures for the same relational R. We say that M' is *iso-preserving* relative to M if for all A and B in $\mathbb{D}_R$ and for

all bijections φ from A to B,

$$M(A) = M(B) \circ \phi \quad \text{implies} \quad M'(A) = M'(B) \circ \phi \, .$$

Let us consider an example. Suppose that the relation R is a weak order such that the range of R, i.e. $\mathbb{D}_R$, consists only of finite sets. Then the function μ with $\mathbb{D}_R$ as range of definition defined by

$$\mu(A) = \beta(R(A))$$

where β is the Borda representation, is a measure for R into $\langle \text{Int}, \geq \rangle$. We call μ the *Borda measure* for R. The Borda measure is structure determined but neither stable nor weakly stable.

This example illustrates the relationship between measures and homomorphic representations. Let us look at it in more detail.

If H is a homomorphic representation of a class containing $\mathbb{E}_R$ (the extension class of R), then we can define a measure M for R in terms of H in the following way: For all A in $\mathbb{D}_R$,

$$M(A) = H(R(A)) \, .$$

On the other hand, if M is a measure for R with respect to Z, then we can define a homomorphic representation H from M in the following way: For all A in $\mathbb{D}_R$,

$$H(R(A)) = M(A) \, .$$

H is thus a homomorphic representation of $\mathbb{E}_R$ into Z. If M is structure determined, then H, which is of course also structure determined, can be extended in a natural way to a homomorphic representation of $\mathbb{C}_R$. If A in $\mathbb{C}_R$ then, according to the definition of $\mathbb{C}_R$, there is B in $\mathbb{D}_R$ such that A and R(B) are isomorphic. Thus there will always be B and φ such that $\phi \in \mathbb{I}(A, R(B))$. Define now the function H^o on $\mathbb{C}_R$ in the following way. For all A in $\mathbb{C}_R$,

$$H^o(A) = H(R(B)) \circ \phi,$$

where $\phi \in \mathbb{I}(A, R(B))$. It is important here that the definition of $H^o(A)$ is independent of the actual choice of B and φ , i.e. the following condition is satisfied: If $\phi' \in \mathbb{I}(A, R(B'))$ then

$$H(R(B)) \circ \phi = H(R(B')) \circ \phi' \, .$$

This is easily proved. H^o is a homomorphic representation of $\mathbb{C}_R$ into Z and for all A in $\mathbb{E}_R$, $H^o(A) = H(A)$. H^o is thus an extension of H to $\mathbb{C}_R$.

A measure M for R has $\mathbb{D}_R$ as its domain. In some contexts it is natural to ask not just what M(A) is but also what M(A) would be if R(A) were A. If M is structure determined then there is a natural answer—given that A in $\mathbb{C}_R$—and that is

$$M(B) \circ \phi,$$

where $\varphi \in I(A,R(B))$. It is easy to see that the actual choice of B and φ is of no importance, i.e. if $\varphi' \in I(A,R(B'))$ then

$$\varphi \circ M(B) = \varphi' \circ M(B') .$$

On the other hand, if M is not structure determined, then it is in general not possible to say what M(A) would be if R(A) were **A**, where **A** is in C_R and R(A)≠**A**. But if M is defined in a special way it may nevertheless be possible. Let us consider this.

A measure is often defined in terms of a homomorphic representation of a considerably larger class than E_R. The Borda measure is an example of this. Suppose M is defined by

$$M(A) = H(R(A))$$

and that H is a homomorphic representation of K, where $E_R \subset K$. Then it is meaningful to talk about what M(A) would be given that R(A)=**A** and **A**∈ K, viz. H(**A**).

Note that

$$\text{For all A in } D_R, \, M(A) = H(R(A))$$

can be written

$$M = H \circ R$$

where ∘ is functional composition, since

$$(H \circ R)(A) = H(R(A)) .$$

That M is defined by M=H∘R will often be written

$$M =_{\text{def}} H \circ R .$$

Suppose M=$_{\text{def}}$ H∘R and M'=$_{\text{def}}$ H'∘R. If H(R(A)) = H'(R(A)) for all A in D_R, then M(A)=M'(A) for all A∈ D_R. Hence, M and M' are equal but not necessarily identical. What M(A) and M'(A) would be if R(A) were **A** need not be the same, since H(**A**) may differ from H'(**A**).

We shall now make a supposition of a special kind: If M=$_{\text{def}}$ H∘R then the domain of H will contain PE_R, i.e. the class of possible extensions of R. Formally this supposition implies a restriction; if we see M and H as function-valued functions and R as a relation-valued function, it is unproblematic that the domain of H does not contain PE_R (but H must contain E_R). But thinking of M as a measure for R and R as a relational, the supposition is very natural. Why? If we define M as H∘R we decide what M(A) is by looking at the value of H for R(A). But we can be mistaken about what R(A) is, and it must be possible to decide what M(A) is given this mistaken value of R(A). Or we might be interested in deciding what M(A) is given the hypothesis that R(A) is **A** (which is actually not the case). But we can decide M(A) given the mistaken or hypothetical value **A** of R(A) only if **A** is a possible value for R, i.e. if **A** is a possible extension. Therefore it is sufficient if H contains $\mathbb{PE}_R$. Thus, if M$=_{def}$HoR then we suppose that the domain of H contains $\mathbb{PE}_R$.

So far we have presupposed that measures for a relational are total in the sense that they are defined for the range of definition of the relational. But there are often reasons to consider even partial measures, i.e. functions that are like measures but only defined for a subset of the range of definition of the relational. Thus, M is a partial measure for R with respect to $\langle Z, \sigma \rangle$ defined for Δ if for all A in Δ, where Δ is a subset of $\mathbb{D}_R$,

$$M(A): A \to Z$$

and for all $a_1, ..., a_n$ in A

$$R(A; a_1, ..., a_n) \text{ iff } \sigma(M(A)(a_1), ..., M(A)(a_n)).$$

For M to be an interesting measure, Δ must be large enough. Δ can for instance be the subset of all finite elements in $\mathbb{D}_R$ or the subset of all elements containing a special element e which is used for defining M(A). (A measure M for a physical aspect is often defined in the following way: M(A) is the homomorphism from R(A) into **Z** such that M(A)(e)=1.)

We end this section with the observation that what has been said about measures for a relational can easily be extended to measures for a relational system. For example, M is a measure for the relational system $\langle R_1, ..., R_k \rangle$ with respect to $\langle Z, \rho_1, ..., \rho_k \rangle$ if for all A in the range of definition of $\langle R_1, ..., R_n \rangle$,

$$M(A): A \to Z$$

and for all j, $1 \leq j \leq k$, and all $a_1, ..., a_j$ in A,

$$R_j(A; a_1, ..., a_j) \text{ iff } \rho_j(M(A)(a_1), ..., M(A)(a_j)).$$

6.5 Numerical measures and representations

What we regard as typical measures are, with few exceptions, numerical in character, i.e. the value of a measure for an object in a domain is usually a number or a k-tuple of numbers. Let us say that the measure M for R with respect to **Y** is a k-*dimensional numerical measure* if $Y \in Re^k$, where k is a positive integer. For all A in $\mathbb{D}_R$ and all a in A, M(A)(a) is then an element in Re^k. If k=1 and M thus a one-dimensional (unidimensional) measure, we also say that M is *real-valued*; for a real-valued measure M, M(A)(a) is thus always a real number.

Let us say that a relational structure $\mathbf{Y}$ is a *k-dimensional numerical structure* if $\mathbf{Y} \subseteq \mathrm{Re}^k$. Hence, if $\mathbf{Y} \subseteq \mathrm{Re}$ then $\mathbf{Y}$ is one-dimensional or unidimensionally numerical structure. A measure M for a relational R with respect to $\mathbf{Y}$ is thus a k-dimensional numerical measure iff $\mathbf{Y}$ is a k-dimensional numerical structure. Let us for convenience say "numerical" instead of "one-dimensional numerical", used both of measures and structures. (As far as structures are concerned this is in accordance with Roberts (1979) p. 51.) If a measure or structure is k-dimensionally numerical with $k \geq 2$, then call it *numerical in the wider sense*. If M is a measure for R with respect to $\mathbf{Y}$ we call $\mathbf{Y}$ the *target region* of M.

Let us look at some simple examples. Suppose R is a weak ordering such that $\mathbb{D}_R$ contains only finite sets. The Borda measure μ for R is a numerical measure for R with respect to the numerical structure $\langle \mathrm{Int}, \leq \rangle$, where Int is the set of integers. Suppose now that for all i, $1 \leq i \leq k$, R_i is a weak ordering with $\mathbb{D}$ as its range of definition, where $\mathbb{D}$ contains only finite sets. Let $\mathbf{R} = \langle R_1,...,R_k \rangle$ and $\mu_{\mathbf{R}} = \langle \mu_1,...,\mu_k \rangle$ where μ_i is the Borda measure for R_i and $\langle \mu_1,...,\mu_k \rangle$ is the restricted product of $\mu_1,...,\mu_k$, i.e.

$$\mu_{\mathbf{R}}(A) = \langle \mu_1,...,\mu_k \rangle(A) = \langle \mu_1(A),...,\mu_k(A) \rangle .$$

Define for all i, $1 \leq i \leq k$, the binary relation $\geq_i$ on $\mathrm{Int}^{(k)}$ in the following way: For all $r, s \in \mathrm{Int}^{(k)}$,

$$r \geq_i s \quad \text{iff} \quad (r)_i \geq (s)_i$$

where $(r)_i$ is the ith component of r and $(s)_i$ is the ith component of s. Since $(\mu_{\mathbf{R}}(A)(x))_i = \mu_i(A)(x)$ it follows that

$$\mu_{\mathbf{R}}(A)(x) \geq_i \mu_{\mathbf{R}}(A)(y) \quad \text{iff} \quad \mu_i(A)(x) \geq \mu_i(A)(y) .$$

It is easy to see now that for all A in $\mathbb{D}$ and all x,y in A,

$$R_i(A)(x,y) \quad \text{iff} \quad \mu_{\mathbf{R}}(A)(x) \geq_i \mu_{\mathbf{R}}(A)(y) .$$

$\mu_{\mathbf{R}}$ (i.e. $\langle \mu_1,...,\mu_k \rangle$) is thus a k-dimensional numerical measure for $\langle R_1,...,R_k \rangle$ with respect to the k-dimensional numerical structure $\langle \mathrm{Int}^{(k)}, \geq_1,...,\geq_k \rangle$. Let us call $\mu_{\mathbf{R}}$ the *k-dimensional Borda measure* for $\mathbf{R}$. Note that

$$\mu_{\mathbf{R}}(A) = \langle \beta(R_1(A)),...,\beta(R_k(A)) \rangle = \langle (\beta \circ R_1)(A)),...,(\beta \circ R_k)(A)) \rangle.$$

In order to simplify notation we introduce the notion of the [Cartesian] product of functions. (See for example Kuratowski, 1972, p.62 or Hu, 1964, p. 13). Let $f : X \rightarrow U$ and $g : Y \rightarrow V$. The function $h : (X \times Y) \rightarrow (U \times V)$ defined by

$$h(x,y) = \langle f(x), g(y) \rangle$$

is called the *product* of f and g. Let us denote h by $f \times g$. Thus

$$(f \times g)(x,y) = \langle f(x), g(y) \rangle.$$

This definition can be straightforwardly generalized to the product of n functions. The product of $f_1,...,f_n$ is often denoted by $\times_i f_i$, $1 \leq i \leq n$. Note the difference between the notion of the product of functions and the notion of the restricted product of functions introduced in section 6.1.

We now see that the k-dimensional Borda measure for $R = \langle R_1,...,R_k \rangle$ is the product of the Borda-measures for $R_i, 1 \leq i \leq n$, i.e.

$$\mu_R = \times_i(\beta o R_i).$$

Define the binary relation σ on $Int^{(k)}$ by

$$\langle r_1,...,r_k \rangle \, \sigma \, \langle s_1,...,s_k \rangle \quad \text{iff} \quad \sum_i r_i \geq \sum_i s_i \,.$$

Let us call the relational R_0 such that μ_R is a measure for R_0 with respect to $\langle Int^{(k)}, \sigma \rangle$ the *Borda amalgamation* of $R_1,...,R_k$. It is of interest in the theory of social choice. (See Gärdenfors, 1973.)

In one-dimensional numerical structures the relation = plays a special role. And in one-dimensional numerical structures which function as target regions for measures $\geq$ is often of predominant importance. It is of interest here what is necessary for the measures of R with respect to Y to have the following property: For all measures M and M' for R with respect to Y and all A in $\mathbb{D}$ and all a and b in A

$$M(A)(a) = M(A)(b) \quad \text{iff} \quad M'(A)(a) = M'(A)(b)\,.$$

6.6 Connections between relationals defined by measures

Measures can be used to express connections between relationals. In this section we shall try to hint how this can be done.

Suppose $\mathbb{D}_R = \mathbb{D}_S = \mathbb{D}$ and that M_R is a measure for R with respect to Y and M_S is a measure for S with respect to Z. Let us further suppose that
(1) for all A in $\mathbb{D}$,

$$M_S(A) = f o M_R(A)\,.$$

This means that for all A in $\mathbb{D}$ and all a in A

$$M_S(A)(a) = f(M_R(A))(a)\,.$$

One could perhaps describe this by saying that there is a functional connection between M_R and M_S. Intuitively this expresses a kind of connection even between R and S; note that for all A in $\mathbb{D}$

$$S(A) = (f o M_R(A))^{-1}[Y]\,.$$

We are interested in whether this connection implies some kind of dependence-relation between R and S and, if this is the case, what kind of dependence.

But (1) above is just one form of a connection between R and S expressed by measures. There are also other forms, for instance

(2) for all A in $\mathbb{D}$ there is a function f_A such that

$$M_S(A) = f_A o \, M_R(A)$$

(3) for all A in $\mathbb{D}$

    (i) $S(A) = (foM_R(A))^{-1}[Z]$ and

    (ii) $(foM_R(A))^{-1}[Z] = (foM'_R(A))^{-1}[Z]$

for all measures M'_R for R with respect to Z.

A connection of another kind between R and S is the following:

(4) M_S is iso-preserving relative to M_R, i.e. if φ is a bijection from A to B, then for all A and B in $\mathbb{D}$,

$$M_R(A) = M_R(B) o \, \varphi \quad \text{implies} \quad M_S(A) = M_S(B) o \, \varphi \, .$$

Let us look at a concrete example of a connection between relationals in terms of measures in order to illustrate the notations introduced here. Suppose that for all i, $1 \leq i \leq k$, R_i is a weak ordering with $\mathbb{D}$ as its range of definition where $\mathbb{D}$ contains only finite sets. Let μ_i be the Borda measure for R_i. As was pointed out in section 6.5, $\mu_R = \langle \mu_1,...,\mu_k \rangle$ is a measure for $\mathbf{R} = \langle R_1,...,R_k \rangle$ with respect to $\mathbf{Y} = = \langle Int^k, \geq_1,...,\geq_k \rangle$. Let f be the function on Re^k into Re defined by

$$f(r_1,...,r_k) = \sum_i r_i \, .$$

And let $\mathbf{Z} = \langle Int, \geq \rangle$. Define a relational R with $\mathbb{D}$ as its range of definition such that for all A in $\mathbb{D}$

$$R_0(A) = (fo\mu_{\mathbf{R}}(A))^{-1}[Z] \, .$$

R_0 is of course binary. Hence,

$$R_0(A)(x,y) \text{ iff } (fo\mu_{\mathbf{R}}(A))(x) \geq (fo\mu_{\mathbf{R}}(A))(y) \text{ iff } f(\mu_{\mathbf{R}}(A)(x)) \geq f(\mu_{\mathbf{R}}(A)(y)) \text{ iff}$$
$$\sum_i \mu_i(A)(x) \geq \sum_i \mu_i(A)(y) \, .$$

R_0 is the Borda amalgamation of $R_1,...,R_k$ as this is defined in the end of section 6.5. The connection between R_0 and $(R_1,...,R_k)$ is of type (1) (see the beginning of this section). Since $\mu_{\mathbf{R}} = \times_i (\beta o R_i)$ (see section 6.5) it follows that

$$R_0(A) = (fo \times_i (\beta o R_i))^{-1}[Z] = (\sum_i (\beta o R_i)(A))^{-1}[Z] \, .$$

Functional connections between measures are of fundamental importance in science. Numerical laws, without which much of modern science could hardly be thinkable, are usually of this kind. In measurement theory the notion (perhaps more precisely the notions) of invariance has (have) been used for studying numerical laws and the topic for this section is closely related to this investigation.

In section 6.1. we proved within the traditional measurement theoretic framework that equality preservation holds iff there is a functional dependence between the corresponding measures. In a quite analogous way we can now prove the following:

Theorem 6.6.1: Suppose R and S are relationals with the same range of definition $\mathbb{D}$. Suppose further that (1) and (2) below holds.
(1) M_R is a real-valued measure for R such that for all $A \in \mathbb{D}$ and all $a, b \in A$, $M_R(A)(a) = M_R(A)(b)$ iff $a \sim_R b; A$.
(2) M_S is a real-valued measure for S such that for all $A \in \mathbb{D}$ and all $a, b \in A$, $M_S(A)(a) = M_S(A)(b)$ iff $a \sim_S b; A$.
Then $S conf R$ iff for all $A \in \mathbb{D}$ there is a function f_A such that
$$M_S(A) = f_A \circ M_R(A)$$
Proof: Let $f_A = \{\langle M_R(A)(a), M_S(A)(a) \rangle \mid a \in A\}$ ◆

The following lemma will be used in the proof of the next two theorems.

Lemma 6.6.2: Suppose M and M' are measures for R with respect to $\mathbb{Z}$ and $A, B \in \mathbb{D}_R$. If $\varphi \in \mathbb{B}i(A,B)$ and $M(A) = M'(B) \circ \varphi$ then $\varphi \in \mathbb{I}(R(A), R(B))$.

The proof of the lemma is straightforward and omitted.
We suppose in the following two theorems that $\mathbb{D}_R = \mathbb{D}_S = \mathbb{D}$.

Theorem 6.6.3: Suppose that M_R is a structure determined measure for R; and M_S is a measure for S. If for all A in $\mathbb{D}$,
$$M_S(A) = f \circ M_R(A)$$
then S is subordinate to R.
Proof: Suppose that $\varphi \in \mathbb{I}(R(A), R(B))$. Since M_R is structure determined $M_R(A) = M_R(B) \circ \varphi$. $M_S(A) = f \circ M_R(A)$ according to the assumption. Thus
$$M_S(A) = f \circ M_R(B) \circ \varphi = M_S(B) \circ \varphi.$$
From 6.6.2 follows $\varphi \in \mathbb{I}(S(A), S(B))$. ◆

Theorem 6.6.4: If for any measure M for R with respect to $\mathbb{Y}$, M_S defined by
$$\text{for every } A \in \mathbb{D}, \quad M_S(A) = f \circ M(A)$$
is a measure for S with respect to $\mathbb{Z}$, then S is subordinate to R.
Proof: Suppose $\varphi \in \mathbb{I}(R(A), R(B))$ and M a measure for R with respect to $\mathbb{Y}$. Let

M* be defined such that M*(A)=M(B)∘φ and M*(X)=M(X) if X≠A. M* is a measure for R with respect to **Y**. Let M_S and M_S* be defined by $M_S(X)$=f∘M(X) and $M_S*(X)$=f∘M*(X) for every X∈ $\mathbb{D}$. Thus M_S and M_S* are measures for S with respect to **Y**. Thus

$$M_S*(A) = f \circ M*(A) = f \circ M(B) \circ \varphi = M_S(B) \circ \varphi$$

According to 6.6.2, φ∈ $\mathbb{I}$(S(A),S(B)). ◆

Note that that 6.6.4 also can be formulated as follows. If for all A in $\mathbb{D}$,

$$S(A) = (f \circ M(A))^{-1}[\mathbf{Z}]$$

for all measures M for R with respect to **Y**, then S is subordinate to R.

PART THREE

FORMAL TREATMENT OF BASIC TOPICS

PART THREE

FORMAL TREATMENT OF BASIC TOPICS

CHAPTER 7

TRANSITIONS BETWEEN SYSTEMS OF RELATIONALS

7.0 Introduction

In this chapter we shall study transitions between relational systems. Transitions were first introduced in section 4.1. We are interested in transitions as a device for studying dependence, and this will determine the form and content of the chapter. In section 7.1 we generalize the notion of finitary system of relationals from section 3.4 so that a system of relationals may consist of infinitely many relationals, and give a short recapitulation of central notions such as range of definition, extension and characteristic class, domain stability etc. The investigation of transitions is begun in section 7.2, where we study relationships between properties of transitions and subordination. In section 7.3 we conduct a similar investigation for correlation rather than subordination.

Section 7.4 deals with the transitions in connection with concatenation of systems of relationals. In section 7.5 we continue the study of significance we began in section 4.9 and transitions are here an important device. Section 7.6 is devoted to monotonicity and stability for transitions and s-functions.

In this chapter a lot of propositions are proved. Those of greatest immediate interest will be called theorems. Propositions which are needed in the context but do not deal with the actual subject matter will be called lemmas.

Elementary results about isomorphisms will be frequently used in this chapter. In section 2.9 there is a collection of such results which will be used without explicit references.

Note that we will often use the expression "relational system" rather than "system of relationals" for convenience. It is therefore important that the reader remember the difference between relational structures and relational systems.

7.1 Relational systems

The notion of a finitary relational system was introduced in section 3.4 We shall now generalize this notion to relational systems in general, i.e. systems not restricted to containing only finite many relationals.

A relational system $\mathbf{R}$ is an ordinal-termed sequence whose terms are relationals with the same range of definition , i.e. $\mathbf{R} = \langle R_i : i < \alpha \rangle$ where R_i is a relational with the same range of definition for all $i < \alpha$. If there is $i < \alpha$ such that $R_i = Q$, then we say that Q is a *term* in $\mathbf{R}$ or is *contained* in $\mathbf{R}$. If α is finite, for example n, we write $\mathbf{R}$ as $\langle R_0,...,R_{n-1} \rangle$. Note the difference between a relational structure and a relational system. A relational structure is an ordered pair $\langle A, \Pi \rangle$ where A is a non-void set and Π is a sequence whose terms are relations on A. A relational system is a sequence whose terms are relationals. An essential point is that we regard a relational system $\mathbf{R}$ by analogy with a relational, i.e. as a function with sets as arguments and relational structures as values. The range of definition of $\mathbf{R}$ is the range of definition of all its terms, and we will write it $\mathbb{D}_{\mathbf{R}}$ or just $\mathbb{D}$. If $\mathbf{R} = \langle R_i : i < \alpha \rangle$ and $A \in \mathbb{D}_{\mathbf{R}}$ then

$$\mathbf{R}(A) = \langle A, (R_i(A))_{i<\alpha} \rangle .$$

If $\mathbf{R}(A) = \langle A, R_i(A) \rangle_{i<\alpha}$ then we call both $\langle A, R_i(A) \rangle_{i<\alpha}$ and $\langle R_i(A) \rangle_{i<\alpha}$ the *extension* of $\mathbf{R}$ on A. When we want to emphasize that it is the one and not the other of these entities we refer to, we call the first the *graph* of R on A and denote it $\underline{\mathbf{R}}(A)$, and call the second the *extension proper* of R on A and denote it $\mathbf{R}''(A)$.

The range of definition of $\mathbf{R}$ is the domain of $\mathbf{R}$ regarded as a function. The arguments of $\mathbf{R}$ will often be called domains for $\mathbf{R}$. The set A is thus a domain for $\mathbf{R}$ iff $A \in \mathbb{D}_{\mathbf{R}}$. To avoid confusion we will avoid "domain" in connection with relationals when we mean "range of definition".

Proposition 7.1.1: If $\mathbf{R} = \langle R_i : i < \alpha \rangle$ then
$$\mathbb{I}(\mathbf{R}(A), \mathbf{R}(B)) = \bigcap_{i<\alpha} \mathbb{I}(R_i(A) R_i(B)).$$
Proof: $\mathbf{R}(A) = \langle A, R_i(A) \rangle_{i<\alpha}$ and $\mathbf{R}(B) = \langle B, R_i(B) \rangle_{i<\alpha}$. Since
$$\mathbb{I}(\langle A, (R_i(A))_{i<\alpha}, \langle B, R_i(B) \rangle_{i<\alpha} \rangle) = \bigcap_{i<\alpha} \mathbb{I}(R_i(A), R_i(B))$$
we have

$$I(R(A),R(B)) = \bigcap_{i<\alpha} I(R_i(A),R_i(B)). \quad \blacklozenge$$

The *extension class* of the relational system R, with the range of definition D_R, is the set $\{R(A) \mid A \in D_R\}$ and we denote it E_R. The *characteristic class* of R is E_R closed under isomorphisms and we denote it C_R. Hence,

$$C_R = \{X \mid \exists A \in D_R : R(A) \cong X\}$$

where $X \cong Y$ means that X and Y are isomorphic. The *restricted characteristic class* of R (earlier in the book called possible extension class) is the class of elements X of C_R such that the domain of X belongs to D_R. We denote this class RC_R, so

$$RC_R = \{X \mid X \in D_R \ \& \ \exists A \in D: R(A) \cong X\} \ .$$

Note that $X \in C_R$ iff $X = \varphi[R(A)]$ where $A \in D_R$ and $\varphi \in Bi(A,X)$.

A relational system R is *domain-stable* if for all $A,B \in D_R$

$$R(A)/B = R(B)/A.$$

R is *weakly domain-stable* if for all $A,B \in D_R$ such that $A \supseteq B$,

$$R(B) = R(A)/B \quad \blacklozenge$$

Proposition 7.1.2: If R is domain-stable then R is weakly domain-stable.
Proof: Suppose $A,B \in D_R$ such that $A \supseteq B$. Since R is domain-stable

$$R(A)/B = R(B)/A.$$

From $A \supseteq B$ follows $R(B)/A = R(B)$. Hence,

$$R(B) = R(A)/B. \quad \blacklozenge$$

Proposition 7.1.3: If $U_R \in D_R$ and R is weakly domain-stable, then for all $A \in D_R$

$$R(A) = R(U_R)/A \ .$$

The proof follows immediately from the definition of weak domain stability.

Proposition 7.1.4: If $U_R \in D_R$ then R is domain-stable iff R is weakly domain-stable.
Proof: Suppose R is weakly domain-stable and that $A,B \in D_R$. Then, according to 7.1.3, $R(A) = R(U_R)/A$ and $R(B) = R(U_R)/B$. From the definition of / (the restriction of a relation to a set) follows (where n is the arity of R)

$$R(A)/B = (R(U_R)/A)/B = (R(U_R) \cap A^n) \cap B^n = R(U_R) \cap (A \cap B)^n =$$
$$R(U_R)/(A \cap B).$$

Analogously we prove that

$$R(B)/A = R(\mathbb{U}_R)/B \cap A .$$

Hence, $R(A)/B = R(B)/A . \quad \blacklozenge$

A set X is *closed under intersection* if for all $A,B \in X$ it holds that $A \cap B \in X$.

Proposition 7.1.5: If $\mathbb{D}_R$ is closed under intersection, then **R** is domain-stable iff **R** is weakly domain-stable.

Proof: Suppose **R** is weakly domain-stable and that $A,B \in \mathbb{D}_R$. Since $\mathbb{D}_R$ is closed under intersection, $A \cap B \in \mathbb{D}_R$.

Hence,

$$R(A \cap B) = R(A)/B \quad \text{and} \quad R(A \cap B) = R(B)/A .$$

This implies that

$$R(A)/B = R(B)/A . \quad \blacklozenge$$

7.2 Transition and subordination

This section contains some elementary results on transitions formulated as a number of propositions, the proofs of which are quite simple. In spite of this they are given here for the sake of completeness. These propositions are used to study transitions between relational systems such that one is subordinate to the other. The main result of this section is that the transition from a relational system **R** to a relational system **S** is a function iff **S** is subordinate to **R**. We start off with definitions of subordination and transition.

Suppose **R** and **S** are relational systems with the same range of definition $\mathbb{D}$. We say that **S** is *subordinate* to **R** if for all $A,B \in \mathbb{D}$

$$\mathbb{I}(S(A),S(B)) \supseteq \mathbb{I}(R(A),R(B)).$$

That S is subordinate to **R** we write for simplicity S*sub*R. We say further that **R** and **S** are *on a par* if R*sub*S and S*sub*R, and we write this R*par*S. Hence, **R** *par* S iff for all $A,B \in \mathbb{D}$

$$\mathbb{I}(R(A),R(B)) = \mathbb{I}(S(A),S(B)).$$

It is obvious that subordination (*sub*) is transitive and that parity (*par*) is an equivalence relation. Hence, *sub* and *par* together constitute a preorder.

As was mentioned earlier, it is often convenient to write $\mathbb{I}(R(A),R(B))$ as $\mathbb{I}_R(A,B)$, and we shall frequently do so.

Suppose **R** and **S** are relational systems with the same range of definition $\mathbb{D}$.

The *transition* from $\mathbf{R}$ to $\mathbf{S}$ is the class

$$\{\langle X,Y \rangle \in \mathbb{C}_{\mathbf{R}} \times \mathbb{C}_{\mathbf{S}} \mid X=Y \ \& \ \exists A \in \mathbb{D}: \exists \varphi \in \mathrm{Bi}(X,A): \varphi \in \mathbb{I}(X,\mathbf{R}(A)) \ \&$$
$$\varphi \in \mathbb{I}(Y,\mathbf{S}(A))\}$$

and we denote it by $\mathbf{R}^{\mathbf{S}}$ We usually write $X\mathbf{R}^{\mathbf{S}}Y$ instead of $\langle X,Y \rangle \in \mathbf{R}^{\mathbf{S}}$. The transition from $\mathbf{R}$ to $\mathbf{S}$, $\mathbf{R}^{\mathbf{S}}$, is to distinguished from the set of functions on A into B. The latter is often denoted B^A, but we shall use $^A B$ instead. The *restricted transition* from $\mathbf{R}$ to $\mathbf{S}$ is the class

$$\{\langle X,Y \rangle \in \mathbb{C}_{\mathbf{R}} \times \mathbb{C}_{\mathbf{S}} \mid X=Y \ \& \ X \in \mathbb{D} \ \& \ \exists A \in \mathbb{D}: \exists \varphi \in \mathrm{Bi}(X,A):$$
$$\varphi \in \mathbb{I}(X,\mathbf{R}(A)) \ \& \ \varphi \in \mathbb{I}(Y,\mathbf{S}(A))\}$$

and we denote it by $\mid \mathbf{R}^{\mathbf{S}} \mid$. Note that

$$X \mid \mathbf{R}^{\mathbf{S}} \mid Y \ \text{iff} \ X\mathbf{R}^{\mathbf{S}}Y \text{ and } X \in \mathbb{D}.$$

In the proofs of the propositions in this section we shall frequently use the elementary results about bijections, isomorphisms, the images of structures under functions and so on, which are collected in chapter 2, without explicit references.

In the rest of this section we suppose that all relational systems under consideration have the same range of definition $\mathbb{D}$.

Proposition 7.2.1: $A\mathbf{R}^{\mathbf{S}}B$ iff there is $C \in \mathbb{D}$ and $\varphi \in \mathrm{Bi}(C,A)$ such that $A=\varphi[\mathbf{R}(C)]$ and $B=\varphi[\mathbf{S}(C)]$.
Proof: According to the definition of $\mathbf{R}^{\mathbf{S}}$ it follows that $A\mathbf{R}^{\mathbf{S}}B$ iff there is $C \in \mathbb{D}$ and $\psi \in \mathrm{Bi}(A,C)$ such that $\psi \in \mathbb{I}(A,\mathbf{R}(C))$ and $\psi \in \mathbb{I}(B,\mathbf{S}(C))$. Let $\varphi = \psi^{-1}$. Hence $\varphi \in \mathrm{Bi}(C,A)$ and $A=\varphi[\mathbf{R}(C)]$ and $B=\varphi[\mathbf{S}(C)]$. $\blacklozenge$

Proposition 7.2.2: If $X \in \mathbb{C}_{\mathbf{R}}$ then there is $Y \in \mathbb{C}_{\mathbf{S}}$ such that $\mathbf{R}^{\mathbf{S}}$. And if $Y \in \mathbb{C}_{\mathbf{S}}$ then there is $X \in \mathbb{C}_{\mathbf{R}}$ such that $\mathbf{R}^{\mathbf{S}}$.
Proof: If $X \in \mathbb{C}_{\mathbf{R}}$ then $X=\varphi[\mathbf{R}(A)]$ where $A \in \mathbb{D}$ and φ a bijection. Let $Y=\varphi[\mathbf{S}(A)]$. Hence, $\varphi^{-1} \in \mathbb{I}(X,\mathbf{R}(A))$ and $\varphi^{-1} \in \mathbb{I}(Y,\mathbf{S}(A))$. From the definition of $\mathbf{R}^{\mathbf{S}}$ follows $X\mathbf{R}^{\mathbf{S}}Y$. The proof of the second part of the proposition is analogous. $\blacklozenge$

Proposition 7.2.2 implies that the transition from $\mathbf{R}$ to $\mathbf{S}$ is a binary relation with $\mathbb{C}_{\mathbf{R}}$ as domain and $\mathbb{C}_{\mathbf{S}}$ as range, i.e. $\mathbf{R}^{\mathbf{S}}$ is a correspondence on $\mathbb{C}_{\mathbf{R}}$ onto $\mathbb{C}_{\mathbf{S}}$.

Proposition 7.2.3: $A\mathbf{R}^{\mathbf{S}}B$ iff $B\mathbf{S}^{\mathbf{R}}A$.
Proof: Suppose $A\mathbf{R}^{\mathbf{S}}B$. Then there exists $C \in \mathbb{D}$ and a bijection φ such that $A=\varphi[\mathbf{R}(C)]$ and $B=\varphi[\mathbf{S}(C)]$. Hence, $\varphi^{-1} \in \mathbb{I}(B,\mathbf{S}(C))$ and $\varphi^{-1} \in \mathbb{I}(A,\mathbf{R}(C))$. From

the definition of S^R follows BS^RA . The proof of the converse is analogous.

Proposition 7.2.4: If AR^SB and $\varphi \in \mathbb{B}i(A,C)$, then $\varphi[A] \, R^S\varphi[B]$.
Proof: Since AR^SB it holds that $A=B$ and there is $D \in \mathbb{D}$ and $\psi \in \mathbb{B}i(D,A)$ such that $\psi \in \mathbb{I}(R(D),A)$ and $\psi \in \mathbb{I}(S(D),B)$. Together with $\varphi o\psi \in \mathbb{B}i(D,C)$ this implies $\varphi o\psi \in \mathbb{I}(R(D),\varphi[A])$ and $\varphi o\psi \in \mathbb{I}(S(D),\varphi[B])$. From the definition of R^S then follows $\varphi[A] \, R^S\varphi[B]$. ♦

Let us define $R^S[A]$ for $A \in \mathbb{C}_R$ as the set of elements in $\mathbb{C}_S$ to which A is related by R^S, thus
$$R^S[A] = \{ X \mid AR^SX \}.$$
Note that $R^S[A]$ consists of exactly one element for every $A \in \mathbb{D}_R$ iff R^S is a function.

Proposition 7.2.5: Suppose $A \in \mathbb{C}_R$ and $\varphi \in \mathbb{B}i(A,B)$. Then
$$R^S[\varphi[A]] = \{ \varphi[X] \mid X \in R^S[A] \}.$$
Proof: Suppose $B \in R^S[A]$. Hence, AR^SB and, according to 7.2.4, $\varphi[A]R^S\varphi[B]$. From this follows immediately $\varphi[B] \in R^S[\varphi[A]]$.

Suppose now $B \in R^S[\varphi[A]]$. Then, according to 7.2.4, $AR^S\varphi^{-1}[B]$ and thus $\varphi^{-1}[B] \in R^S[A]$. From this follows $B \in \{ \varphi[X] \mid X \in R^S[A] \}$. ♦

Proposition 7.2.6: If $A \in \mathbb{C}_R$ then
$$R^S[A] = \{ \varphi[S(X)] \mid X \in \mathbb{D} \ \& \ \varphi \in \mathbb{I}(R(X),A) \} .$$
Proof: (1) Suppose $B \in R^S[A]$. From the definition of R^S follows $A=B$ and there is $C \in \mathbb{D}$ and $\psi \in \mathbb{B}i(A,C)$ such that $\psi \in \mathbb{I}(A,R(C))$ and $\psi \in \mathbb{I}(B,S(C))$. Hence, $\psi^{-1} \in \mathbb{I}(R(C),A)$ and $B=\psi^{-1}[S(C)]$, i.e.
$$B \in \{ \varphi[S(X)] \mid X \in \mathbb{D} \ \& \ \varphi \in \mathbb{I}(R(X),A) \} .$$
(2) Suppose $X \in \mathbb{D}$ and $\varphi \in \mathbb{I}(R(X),A)$. Then $\varphi[R(X)]=A$. Since $R(X)R^SS(X)$ it follows from proposition 7.2.4 that $\varphi[R(X)] \, R^S\varphi[S(X)]$. Hence, $AR^S\varphi[S(X)]$, i.e. $\varphi[S(X)] \in R^S[A]$. ♦

If $A \in \mathbb{D}$ let us for simplicity use $R^S[A]$ as short for $R^S[R(A)]$. It is obvious that for every $A \in \mathbb{D}$, $S(A) \in R^S[A]$.

Corollary 7.2.7: If $A \in \mathbb{D}$ then
$$\mathbf{R}^S[A] = \{\varphi[S(X)] \mid X \in \mathbb{D} \ \& \ \varphi \in \mathbb{I}_\mathbf{R}(X,A)\} .$$

Proposition 7.2.8: $\mathbf{R}^S$ is a function iff for every $A \in \mathbb{D}$, $\mathbf{R}^S[A]=\{S(A)\}$.
Proof: (1) Suppose $\mathbf{R}^S$ is a function. Hence, $\mathbf{R}^S[A]$ is a singelton, (i.e. consists of only one element). Since $S(A) \in \mathbf{R}^S[A]$ it follows that $\mathbf{R}^S[A]=\{S(A)\}$.
(2) Suppose $\mathbf{R}^S[A]=\{S(A)\}$ for every $A \in \mathbb{D}$. Let $B \in \mathbb{C}_\mathbf{R}$. Then there is $C \in \mathbb{D}$ and $\varphi \in \mathbb{B}i(C,B)$ such that $\varphi[R(C)]=B$. From $\mathbf{R}^S[C]=\{S(C)\}$ it follows according to proposition 7.2.5, that
$$\mathbf{R}^S[\varphi[R(C)]] = \{\varphi[S(C)]\} .$$
Hence, $\mathbf{R}^S[B]=\{\varphi[S(C)]\}$, i.e. $\mathbf{R}^S[B]$ is a singelton. This implies that $\mathbf{R}^S$ is a function. $\blacklozenge$

We shall use $\mathbf{R}^S$ to study the strength of the dependence between $\mathbf{R}$ and S. In chapter 4 we showed that the transition from R to S is a function iff $S sub B$. We shall prove this for relational systems.

Theorem 7.2.9: $\mathbf{R}^S$ is a function iff $S sub \mathbf{R}$.
Proof: (1) Suppose $\mathbf{R}^S$ is a function. Let $\varphi \in \mathbb{I}_\mathbf{R}(B,A)$. Then $\varphi[S(B)] \in \mathbf{R}^S[A]$ according to 7.2.7. And 7.2.8 implies $\varphi[S(B)]=S(A)$. Hence $\varphi \in \mathbb{I}_S(B,A)$. We have thus proved $S sub \mathbf{R}$.
(2) Suppose $S sub \mathbf{R}$. Let $A \in \mathbb{D}$. If $B \in \mathbf{R}^S[A]$ then, according to 7.2.7, $B=\varphi[S(X)]$, where $X=\mathbb{D}$ and $\varphi \in \mathbb{I}_\mathbf{R}(X,A)$. $\varphi \in \mathbb{I}_S(X,A)$ since $\mathbb{I}_S(X,A) \supseteq \mathbb{I}_\mathbf{R}(X,A)$. Hence, $B=\varphi[S(X)]=S(A)$, which implies that $\mathbf{R}^S[A]$ is a singelton. From 7.2.8 it then follows that $\mathbf{R}^S$ is a function. $\blacklozenge$

Lemma 7.2.10: $\mathbf{R}^S$ is bijective (i.e. one-to-one) iff both $\mathbf{R}^S$ and $S^\mathbf{R}$ are functions.
Proof: (1) Suppose $\mathbf{R}^S$ is bijective. Then $\mathbf{R}^S$ is of course a function. Suppose further that $S^\mathbf{R}$ is not a function i.e. there is $A \in \mathbb{C}_S$ and $B,B' \in \mathbb{D}_S$ where $B \neq B'$ such that $A S^\mathbf{R} B$ and $A S^\mathbf{R} B'$. Then, according to 7.2.3, $B \mathbf{R}^S A$ and $B' \mathbf{R}^S A$. Hence, $\mathbf{R}^S$ is not bijective, which implies a contradiction. We have thus proved that $S^\mathbf{R}$ is a function.
(2) Suppose that $\mathbf{R}^S$ and $S^\mathbf{R}$ are functions but that $\mathbf{R}^S$ is not bijective. Then there exist $A,A' \in \mathbb{C}_\mathbf{R}$ and $B \in \mathbb{C}_S$, where $A \neq A'$, such that $A \mathbf{R}^S B$ and $A' \mathbf{R}^S B$.

Hence, according to 7.2.3, BS^RA and BS^RA', which implies that S^R is not a function. We have thus proved a contradiction, and therefore R^S is bijective. ♦

Theorem 7.2.11: R^S is bijective iff $RparS$.
Proof: (1) Suppose R^S is bijective. According to 7.2.10 R^S and S^R are functions. From 7.2.9 then follows $RsubS$ and $SsubR$ and thus $RparS$.
(2) Suppose $RparS$. Then it follows that $RsubS$ and $SsubR$. From 7.2.9 it follows that R^S and S^R are functions, and thus R^S bijective according to 7.2.10. ♦

Reciprocal subordination, i.e. parity, is a strong form of that kind of dependence between relational systems. The transition between two relational systems is as "small" (restricted) as possible when it is a bijection and then the relational systems are on a par. Let us now look at transitions between relational systems which are as "large" (unrestricted) as possible.

7.3 Transitions and uncorrelation

In this section we study the interplay between subordination and correlation by means of transitions. The main result of this section is that parity is a congruence relation for both correlation and uncorrelation.

In chapter 4 we introduced the notion of uncorrelation and we generalize it here to relational systems. Suppose R and S are relational systems with the same range of definition. S is *uncorrelated* with R iff
$$R^S = \{\langle X,Y\rangle \in \mathbb{C}_R \times \mathbb{C}_S \mid X=Y\} .$$
If R and S have the same range of definition and S is not uncorrelated with R, then R is correlated with S. Since it holds generally that
$$R^S \subseteq \{\langle X,Y\rangle \in \mathbb{C}_R \times \mathbb{C}_S \mid X=Y\},$$
S is *correlated* with R iff
$$R^S \subset \{\langle X,Y\rangle \in \mathbb{C}_R \times \mathbb{C}_S \mid X=Y\}.$$
We let *uncorr* denote the relation "is uncorrelated with" and *corr* the relation "is correlated with". Note that $RcorrS$ iff not $SuncorrR$.

Proposition 7.3.1: If $SuncorrR$ then $RuncorrS$.
Proof: Suppose $X \in \mathbb{C}_S$, $Y \in \mathbb{C}_R$ and $X=Y$. Since S is uncorrelated with R then $Y R^S X$. Hence, XS^RY according to 7.2.3. We have thus proved that
$$S^R = \{\langle X,Y\rangle \in \mathbb{C}_S \times \mathbb{C}_R \mid X=Y\}. ♦$$

Proposition 7.3.2: If $ScorrR$ then $RcorrS$.

Proof: Since S is correlated with R there is $A \in \mathbb{C}_R$, $B \in \mathbb{C}_S$ such that $A=B$ but $\neg AR^SB$. According to 7.2.3, $\neg BS^RA$. Hence,

$$S^R \subset \{\langle X,Y \rangle \in \mathbb{C}_S \times \mathbb{C}_R \mid X=Y\}. \quad \blacklozenge$$

Proposition 7.3.3: $SuncorrR$ iff whenever $A \in \mathbb{C}_R$, $B \in \mathbb{C}_S$ and $A=B$ then there is $C \in \mathbb{D}$ and $\varphi \in \mathbb{B}i(C,A)$ such that

$$A=\varphi[R(C)] \quad \text{and} \quad B=\varphi[S(C)].$$

The proof follows immediately from the definition of uncorr and of R^S. 7.3.1 and 7.3.2 implies that both *uncorr* and *corr* are symmetric. We can therefore say that R and S are [un]correlated instead of saying that R is [un]correlated with S. As we shall see at the end of this section none of *corr* and *uncorr* are reflexive or transitive.

We shall now study the interconnection between *uncorr, corr* and *sub*. First some preliminaries.

The transition R^S from R to S is a correspondence from $\mathbb{C}_R$ to $\mathbb{C}_S$. Hence, R^S relates structures. Since AR^SB implies $A=B$ two corresponding structures have the same domain. These observations motivates the following definition. Suppose each of K and L is a set of similar structures. γ is an *s-correspondence* ("s" for structure) from K to L if $\gamma \subseteq K \times L$ and $X\gamma Y$ implies $X=Y$. If the domain of γ is K then γ is on K. If the range of γ is L then γ is onto L. R^S is thus an s-correspondence on $\mathbb{C}_R$ onto $\mathbb{C}_S$.

Since s-correspondences are binary relations we can talk about (1) the relative product of two s-correspondences and (2) the inverse of an s-correspondence. We denote the relative product of two s-correspondences γ and δ by $\gamma \mid \delta$ and the inverse of an s-correspondence γ by γ^{-1}. Hence,

$$\gamma \mid \delta = \{\langle x,y \rangle \mid \exists z: x\gamma z \ \& \ z\delta y\}$$
$$\gamma^{-1} = \{\langle x,y \rangle \mid y\gamma x\}.$$

As was pointed out in section 2.1, if ρ is a binary relation then the domain of ρ is denoted by $Do\rho$ and the range of ρ is denoted by $Rg\rho$. Note that $Do\gamma = Rg\gamma^{-1}$ and $Do\gamma^{-1}=Rg\gamma$. It is a well-known fact that

$$(\gamma \mid \delta) \mid \varepsilon = \gamma \mid (\delta \mid \varepsilon)$$

i.e. relative product is associative, and that $(\gamma \mid \delta)^{-1}=\delta^{-1} \mid \gamma^{-1}$.

The following lemma states two simple results on correspondences in general.

Lemma 7.3.4: (i) If γ is a correspondence on X onto Y then γ^{-1} is a correspondence on Y onto X.
(ii) If γ is a correspondence on X onto Y and δ is a correspondence on Y onto Z, then $\gamma \mid \delta$ is a correspondence on X onto Z.

Lemma 7.3.5: The inverse of an s-correspondence is an s-correspondence, and the relative product of two s-correspondences is an s-correspondence.

We suppose in the rest of this section—even if we do not always state it—that all relational systems under considerations have the same range of definition $\mathbb{D}$.

Proposition 7.3.6: $(R^S)^{-1} = S^R$.
The proposition follows immediately from the definition of inverse and 7.2.3.

Proposition 7.3.7: $Q^R \mid R^S$ is an s-correspondence of $\mathbb{C}_Q$ onto $\mathbb{C}_S$.
The proposition follows from 7.3.4 and 7.3.5.

Proposition 7.3.8: $Q^R \mid R^S \supseteq Q^S$.
Proof: Suppose AQ^SB. Then there is $C \in \mathbb{D}$ and a bijection φ such that $A = \varphi[Q(C)]$ and $B = \varphi[S(C)]$. From 7.2.4 it follows that
$$\varphi[Q(C)]Q^R\varphi[R(C)] \text{ and } \varphi[R(C)]R^S\varphi[S(C)],$$
i.e. AQ^RD and DR^SB where $D = \varphi[R(C)]$. Hence, $AQ^R \mid R^SB$. $\blacklozenge$

The set inclusion in 7.3.8 can not generally be replaced by equality, which we shall show at the end of this section. However, under some special conditions we can prove that $Q^R \mid R^S = Q^S$.

Proposition 7.3.9: If R^S is a function, i.e. if S *sub* R, then
$$Q^R \mid R^S = Q^S$$
Proof: Since $Q^R \mid R^S \supseteq Q^S$ it is sufficient according to 7.3.8, to prove that
$$Q^R \mid R^S \subseteq Q^S .$$
Suppose $AQ^R \mid R^SB$. Then there is C such that
$$AQ^RC \text{ and } CR^SB .$$

From this it follows that there is $D,E \in \mathbb{D}$ and $\varphi \in \mathrm{Bi}(D,A)$ and $\psi \in \mathrm{Bi}(E,B)$ such that

$$A = \varphi[Q(D)] \, , \quad C = \varphi[R(D)] \quad \& \quad C = \psi[R(E)] \, , \quad B = \psi[S(E)] \, .$$

Since

$$\varphi[R(D)] \mathbf{R}^\mathbf{S} \varphi[S(D)] \quad \& \quad \psi[R(E)] \mathbf{R}^\mathbf{S} \psi[S(E)]$$

and $\mathbf{R}^\mathbf{S}$ is a function it follows that $\varphi[S(D)] = \psi[S(E)] = \mathbf{B}$. $\varphi[Q(D)] \mathbf{Q}^\mathbf{S} \varphi[S(D)]$ thus implies $A\mathbf{Q}^\mathbf{S}\mathbf{B}$. ♦

Proposition 7.3.10: If $\mathbf{R}^\mathbf{Q}$ is a function, i.e. if $Q \mathit{sub} R$, then
$$\mathbf{Q}^\mathbf{R} | \mathbf{R}^\mathbf{S} = \mathbf{Q}^\mathbf{S} \, .$$
Proof: Since $\mathbf{R}^\mathbf{Q}$ is a function it follows from 7.3.9 that
$$\mathbf{S}^\mathbf{R} | \mathbf{R}^\mathbf{Q} = \mathbf{S}^\mathbf{Q} \, .$$
From this and 7.3.6 it follows that
$$\mathbf{Q}^\mathbf{S} = (\mathbf{S}^\mathbf{Q})^{-1} = (\mathbf{S}^\mathbf{R} | \mathbf{R}^\mathbf{Q})^{-1} = (\mathbf{R}^\mathbf{Q})^{-1} | (\mathbf{S}^\mathbf{R})^{-1} = \mathbf{Q}^\mathbf{R} | \mathbf{R}^\mathbf{S}. \text{ ♦}$$

We now introduce a notion applicable to s-correspondences and use it in the study of uncorrelation.

An s-correspondence γ is *sweeping* if for every $X \in \mathrm{Do}\gamma$ and $Y \in \mathrm{Rg}\gamma$, $X=Y$ implies $X\gamma Y$. $\mathbf{R}^\mathbf{S}$ is an s-correspondence on $\mathbb{C}_\mathbf{R}$ onto $\mathbb{C}_\mathbf{S}$, so $\mathbf{R}^\mathbf{S}$ is sweeping iff for every $A \in \mathbb{C}_\mathbf{R}$ and $B \in \mathbb{C}_\mathbf{S}$, $A=B$ implies $A\mathbf{R}^\mathbf{S}B$.

Our interest in the notion of sweeping s-correspondence stems from the following observation.

Proposition 7.3.11: $\mathbf{R}^\mathbf{S}$ is sweeping iff R *uncorr* S .
Proof: (1) Suppose $\mathbf{R}^\mathbf{S}$ is sweeping. Suppose further that $\langle A,B \rangle \in \mathbb{C}_\mathbf{R} \times \mathbb{C}_\mathbf{S}$ and $A=B$. According to the definition of sweeping it holds that $A\mathbf{R}^\mathbf{S}B$. Hence,
$$\mathbf{R}^\mathbf{S} = \{ \langle X,Y \rangle \in \mathbb{C}_\mathbf{R} \times \mathbb{C}_\mathbf{S} \mid X=Y \} \, .$$
R and S are thus uncorrelated.
(2) Suppose that R and S are uncorrelated. Suppose further that $A \in \mathbb{C}_\mathbf{R}$ and $B \in \mathbb{C}_\mathbf{S}$ and $A=B$. According to the definition of uncorrelation it holds that $A\mathbf{R}^\mathbf{S}B$. Hence, $\mathbf{R}^\mathbf{S}$ is sweeping. ♦

Lemma 7.3.12: Suppose $\mathrm{Do}\gamma \subseteq \mathrm{Do}\delta$, $\mathrm{Rg}\gamma \subseteq \mathrm{Rg}\delta$ and $\delta \subseteq \gamma$. If γ is not sweeping, then δ is not sweeping.
Proof: If γ is not sweeping then there is $A \in \mathrm{Do}\gamma$ and $B \in \mathrm{Rg}\gamma$ such that $A=B$ but

$\neg A\gamma B$. Suppose $\delta \subseteq \gamma$. Then $\neg A\delta B$. Since $A\in Do\delta$ and $B\in Rg\gamma$ δ is not sweeping. ◆

Lemma 7.3.13: γ is sweeping iff γ^1 is sweeping.
Proof: Suppose γ is sweeping. Suppose further that $A\in Do\gamma^1$ and $B\in Rg\gamma^1$ and $A=B$. Then $B\gamma A$ since γ is sweeping. Hence $A\gamma^1 B$, so γ^1 is sweeping. The proof that γ is sweeping if γ^1 is sweeping is similar. ◆

For convenience we regard $\varnothing$ as an s-correspondence on $\varnothing$ onto $\varnothing$. $\varnothing$ is, according to the definition of sweeping, a sweeping s-correspondence on $\varnothing$ onto $\varnothing$. Hence, if γ and δ are s-correspondences and $Rg\gamma \cap Do\delta = \varnothing$ then $\gamma | \delta$ is sweeping since $\gamma | \delta = \varnothing$.

Lemma 7.3.14: If γ or δ is sweeping, then $\gamma | \delta$ is sweeping.
Proof: (1) Suppose γ is sweeping. If $\gamma | \delta = \varnothing$ then $\gamma | \delta$ is sweeping. Suppose therefore that $A\in Do(\gamma | \delta)$, $B\in Rg(\gamma | \delta)$ and $A=B$. Then there is $C\in Do(\gamma | \delta)$ such that $C(\gamma | \delta)B$. Hence, there is $D\in Rg\gamma \cap Do\delta$ such that $C\gamma D$ and $D\gamma B$. Since $A=B$ and $D=B$ it holds that $A=D$ and since γ is sweeping $A\gamma D$. We thus have $A\gamma D$ and $D\delta B$, which implies $A(\gamma | \delta)B$. Hence, $\gamma | \delta$ is sweeping.
(2) Suppose now that δ is sweeping. If $\gamma | \delta = \varnothing$ then $\gamma | \delta$ is sweeping. Suppose therefore that $A\in Do(\gamma | \delta)$ and $B\in Rg(\gamma | \delta)$ and $A=B$. Then there is $C\in Rg(\gamma | \delta)$ such that $A(\gamma | \delta)C$. Hence, there is $D\in Rg\gamma \cap Do\delta$ such that $A\gamma D$ and $D\delta C$. Since $A=B$ and $A=D$ it holds that $D=B$ and since δ is total $D\delta B$. ◆

Lemma 7.3.15: Suppose that γ and δ are s-correspondences and $Rg\gamma=Do\delta$. If γ is not sweeping and δ^{-1} is a function, then $\gamma | \delta$ is not sweeping.
Proof: Since γ is not sweeping there is $A\in Do\gamma$ and $B\in Rg\gamma$ such that $A=B$ but $\neg A\gamma B$. Since $Rg\gamma=Do\delta$ there is $C\in Rg\delta$ such that $B\delta C$. Suppose now that $A(\gamma | \delta)C$. Then there is $D\in Rg\gamma$ such that $A\gamma D$ and $D\delta C$. We thus have $C\delta^{-1}B$ and $C\delta^{-1}D$, and since δ^{-1} is a function this implies $B=D$. Hence, $A\gamma B$, which contradicts the assumption $\neg A\gamma B$. We have thus proved that $\neg A\gamma | \delta C$. Since $A=B=C$ this shows that $\gamma | \delta$ is not sweeping. ◆

Corollary 7.3.16: If Q^R is not sweeping and S^R is a function, then $Q^R | R^S$ and Q^S are not sweeping.

Lemma 7.3.17: Suppose that γ and δ are s-correspondences and Rgγ=Doδ. If γ is a function and δ is not sweeping, then $\gamma | \delta$ is not sweeping.

Proof: The proposition can be proved analogously to 7.3.15. But it can also be proved in the following way using 7.3.15. Since δ is not sweeping, δ^{-1} is not sweeping according to 7.3.13. Since γ is a function and Doγ^{-1}=Rgδ^{-1} we can apply 7.3.15, from which follows that $\delta^{-1} | \gamma^{-1}$ is not sweeping. Hence, according to 7.3.13, $(\delta^{-1} | \gamma^{-1})^{-1}$ is not sweeping. Since $(\delta^{-1} | \gamma^{-1})^{-1}=\gamma | \delta$ it follows that $\gamma | \delta$ is sweeping. ◆

Corollary 7.3.18: If Q^R is a function and R^S is not sweeping, then $Q^R | R^S$ and Q^S is not sweeping.

Proposition 7.3.19: If Q*corr*R and R*sub*S , then Q*corr*S .

Proof: Q*corr*R implies that Q^R is not sweeping. R*sub*S implies that S^R is a function. Hence, according to 7.3.16, $Q^R | R^S$ is not sweeping. According to 7.3.8 $Q^R | R^S \supseteq Q^S$. Hence, according to 7.3.12, Q^S is not sweeping, i.e. Q*corr*S according to 7.3.11. ◆

Proposition 7.3.20: If R*corr*S and R*sub*Q, then Q*corr*S.

Proof: R*corr*S implies S*corr*R, so we have S*corr*R and R*sub*Q. From this it follows according to 7.3.19 that S*corr*Q. Hence, Q*corr*S. ◆

Theorem 7.3.21: Suppose R*uncorr*S, Q*sub*R and T*sub*S. Then Q*uncorr*T.

Proof: Let us make the assumption in the theorem but suppose that Q*corr*T. Then it holds that Q*corr*T and T*sub*S. From this follows according to 7.3.19 that Q*corr*S. Together with Q*sub*R this implies according to 7.3.20 that R*corr*S. This contradicts R*uncorr*S. Hence, Q*uncorr*T. ◆

Theorem 7.3.22: Suppose R*corr*S, R*sub*Q and S*sub*T. Then Q*corr*T.

Proof: Let us make the assumption in the theorem but suppose that Q*uncorr*T. We then have Q*uncorr*T, R*sub*Q and S*sub*T. Application of 7.3.21 gives R*uncorr*S. This contradicts R*corr*S. Hence, Q*corr*T. ◆

Let us say that R *is superior to* S if S is subordinate to R. We use "R*sup*S" as an abbreviation for "S*sub*R" and use *sup* to denote the relation "is superior to". *sup* is thus the inverse of *sub*. With the help of *sup* we can formulate 7.3.21 in the following way:

$$\mathbf{Q}sub\mathbf{R}uncorr\mathbf{S}sup\mathbf{T} \implies \mathbf{Q}uncorr\mathbf{T}.$$

And 7.3.22 can be formulated as follows:

$$\mathbf{Q}sup\mathbf{R}corr\mathbf{S}sub\mathbf{T} \implies \mathbf{Q}corr\mathbf{T}.$$

Theorem 7.3.23: Suppose $\mathbf{Q}par\mathbf{R}$ and $\mathbf{S}par\mathbf{T}$. Then $\mathbf{R}uncorr\mathbf{S}$ implies $\mathbf{Q}uncorr\mathbf{T}$ and $\mathbf{R}corr\mathbf{S}$ implies $\mathbf{Q}corr\mathbf{T}$.

Proof: $\mathbf{Q}par\mathbf{R}$ is equivalent to $\mathbf{Q}sub\mathbf{R}$ and $\mathbf{Q}sup\mathbf{R}$. And $\mathbf{S}par\mathbf{T}$ is equivalent to $\mathbf{S}sub\mathbf{T}$ and $\mathbf{S}sup\mathbf{T}$. Then apply 7.3.21 and 7.3.22. ◆

Note that 7.3.23 means that *par* is a congruence relation for *corr* and *uncorr*. We can therefore define relations <u>*corr*</u> and <u>*uncorr*</u> for equivalent classes of relational systems with respect to parity according to the rules

$$[\mathbf{R}]\underline{uncorr}[\mathbf{S}] \quad \text{iff} \quad \mathbf{R}uncorr\mathbf{S}$$

$$[\mathbf{R}]\underline{corr}[\mathbf{S}] \quad \text{iff} \quad \mathbf{R}corr\mathbf{S}$$

where $[\mathbf{R}]$ is the equivalence class of $\mathbf{R}$ with modulo parity in the set of all relational systems with $\mathbb{D}$ as range of definition.

We end this section by studying (1) the transition from a relational system to itself and (2) relational systems with a degenerate characteristic class. The results will be applied to some questions about subordination and uncorrelation.

Proposition 7.3.24: $\mathbf{R}^\mathbf{R} = \{\langle X,X \rangle \mid X \in \mathbb{C}_\mathbf{R}\}$

Proof: (1) Suppose $X \in \mathbb{C}_\mathbf{R}$. Then there is $A \in \mathbb{D}$ and $\varphi \in \mathrm{Bi}(A,X)$ such that $X = \varphi[\mathbf{R}(A)]$. From this and 7.2.1 it follows that $X\mathbf{R}^\mathbf{R}X$.

(2) Suppose $X\mathbf{R}^\mathbf{R}X$. Then, according to 7.2.1, there is $A \in \mathbb{D}$ and $\varphi \in \mathrm{Bi}(A,X)$ such that $X = \varphi[\mathbf{R}(A)]$ and $Y = \varphi[\mathbf{R}(A)]$. Hence $X = Y$. Obviously $X \in \mathbb{C}_\mathbf{R}$. ◆

Proposition 7.3.25: If not $\mathbf{R}sub\mathbf{S}$, then $\mathbf{R}^\mathbf{S}| \mathbf{S}^\mathbf{R} \supset \mathbf{R}^\mathbf{R}$.

Proof: From 7.3.8 it follows that $\mathbf{R}^\mathbf{S}| \mathbf{S}^\mathbf{R} \neq \mathbf{R}^\mathbf{R}$, so it is sufficient to prove that $\mathbf{R}^\mathbf{S}| \mathbf{S}^\mathbf{R} \neq \mathbf{R}^\mathbf{R}$. If not $\mathbf{R}sub\mathbf{S}$ then $\mathbf{S}^\mathbf{R}$ is not a function. There is therefore $X \in \mathbb{C}_\mathbf{S}$ such that $X\mathbf{S}^\mathbf{R}Y$ and $X\mathbf{S}^\mathbf{R}Y'$ and $Y \neq Y'$. Hence, $Y\mathbf{R}^\mathbf{S}X$ and $X\mathbf{S}^\mathbf{R}Y'$, which implies that $Y\mathbf{R}^\mathbf{S}| \mathbf{S}^\mathbf{R}Y'$. But according to 7.3.24 not $Y\mathbf{R}^\mathbf{R}Y'$, so $\mathbf{R}^\mathbf{S}| \mathbf{S}^\mathbf{R} \neq \mathbf{R}^\mathbf{R}$. ◆

Proposition 7.3.25 shows that the set inclusion in 7.3.8 can not in general be replaced by equality.

Let us say that the characteristic class of $\mathbf{R}$, i.e. $\mathbb{C}_{\mathbf{R}}$, is *degenerate* if whenever $X,Y \in \mathbb{C}_{\mathbf{R}}$ and $X \equiv Y$ then $X = Y$. "Ordinary" relational systems have a non-degenerate characteristic class. But there are relationals with a degenerate characteristic class. The n-ary universal relational V defined by $V(A) = A^n$ for all $A \in \mathbb{D}$ and the n-ary empty relation $\varnothing$ defined by $\varnothing(A) = \varnothing$ for all A in $\mathbb{D}$ both have a degenerate characteristic class.

Proposition 7.3.26: If $\mathbb{C}_{\mathbf{R}}$ is degenerate, then $\mathbf{R}sub\mathbf{S}$ for every relational system S with $\mathbb{D}$ as range of definition.
Proof: Suppose $XS^{\mathbf{R}}Y$ and $XS^{\mathbf{R}}Y'$. Then $X \equiv Y$ and $X \equiv Y'$ and thus $Y \equiv Y$. Since $\mathbb{C}_{\mathbf{R}}$ is degenerate it follows that $Y = Y'$. Hence $S^{\mathbf{R}}$ is a function, which implies that $\mathbf{R}sub\mathbf{S}$ according to 7.2.9. ♦

Proposition 7.3.27: $\mathbf{R}uncorr\mathbf{R}$ iff $\mathbb{C}_{\mathbf{R}}$ is degenerate.
Proof: From the definition of *uncorr* and 7.3.24 follows that $\mathbf{R}uncorr\mathbf{R}$ iff
$$\{\langle X,Y \rangle \in \mathbb{C}_{\mathbf{R}} \times \mathbb{C}_{\mathbf{R}} \mid X \equiv Y\} = \{\langle X,X \rangle \mid X \in \mathbb{C}_{\mathbf{R}}\}.$$
which holds iff $\mathbb{C}_{\mathbf{R}}$ is degenerate. ♦

Corollary 7.3.28: $\mathbf{R}corr\mathbf{R}$ iff $\mathbb{C}_{\mathbf{R}}$ is non-degenerate.

Note that if $\mathbb{C}_{\mathbf{R}}$ is degenerate then $\mathbf{R}sub\mathbf{R}$ and $\mathbf{R}uncorr\mathbf{R}$.

7.3.27 shows that *uncorr* is neither reflexive nor irreflexive and 7.3.28 shows that the same holds for *corr*. But if we restrict ourselves to relational systems with non-degenerate characteristic class then *uncorr* is irreflexive while *corr* is reflexive. If $\mathbf{R}uncorr\mathbf{S}$ and $\mathbb{C}_{\mathbf{R}}$ is non-degenerate, then
$$\mathbf{R}uncorr\mathbf{S} \ \& \ \mathbf{S}uncorr\mathbf{R} \ \& \ \mathbf{R}corr\mathbf{R}$$
so *uncorr* is not transitive. We shall return to the question whether *corr* is transitive.

Proposition 7.3.29: If $\mathbb{C}_{\mathbf{R}}$ is degenerate, then $\mathbf{R}uncorr\mathbf{S}$ for all relational systems S (with $\mathbb{D}$ as its range of definition).
Proof: Suppose that $\mathbb{C}_{\mathbf{R}}$ is degenerate and that $\mathbf{R}corr\mathbf{S}$. Then
$$\mathbf{R}^{\mathbf{S}} \subset \{\langle X,Y \rangle \in \mathbb{C}_{\mathbf{R}} \times \mathbb{C}_{\mathbf{S}} \mid X \equiv Y\}$$

so there are $X_0 \in \mathbb{C}_\mathbf{R}$ and $Y_0 \in \mathbb{C}_\mathbf{S}$ such that $X_0 = Y_0$ but not $X_0 \mathbf{R}^\mathbf{S} Y_0$. According to 7.2.2 there is $X_1 \in \mathbb{C}_\mathbf{R}$ such that $X_1 \mathbf{R}^\mathbf{S} Y_0$. Hence $X_1 = Y_0$ and from $X_0 = Y_0$ follows $X_1 = X_0$. Since $\mathbb{C}_\mathbf{R}$ is degenerate it follows that $X_1 = X_0$. $X_1 \mathbf{R}^\mathbf{S} Y_0$ therefore contradicts $X_0 \mathbf{R}^\mathbf{S} Y_0$. ♦

Note that it does not hold generally that *sub* implies *corr*. However, under a general assumption it does.

Proposition 7.3.30: If $\mathbb{C}_\mathbf{R}$ is non-degenerate, then $\mathbf{R}\,subS$ implies $\mathbf{R}\,corrS$.
Proof: Suppose that $\mathbb{C}_\mathbf{R}$ is non-degenerate and that $\mathbf{R}\,uncorrS$. Then there is $Y, Y' \in \mathbb{C}_\mathbf{R}$ such that $Y = Y'$ and $Y \neq Y'$, and there is $X \in \mathbb{C}_\mathbf{R}$ such that
$$X S^\mathbf{R} Y \text{ and } X S^\mathbf{R} Y'.$$
Hence, $\mathbf{S}^\mathbf{R}$ is not a function, i.e. not $\mathbf{R}\,subS$. ♦

7.4 Concatenation and transition

Suppose that $\mathbf{R}$ and $\mathbf{S}$ are relational systems with the same range of definition $\mathbb{D}$. Then both $\mathbf{R}$ and $\mathbf{S}$ are sequences whose terms are relationals; $\mathbf{R} = \langle R_i : i < \alpha \rangle$ and $\mathbf{S} = \langle S_i : i < \beta \rangle$. With $\mathbf{RS}$, the concatenation of $\mathbf{R}$ and $\mathbf{S}$, we mean the concatenation of the sequences $\mathbf{R}$ and $\mathbf{S}$ (see section 2.8), i.e. the relational system $\mathbf{Q} = \langle Q_i : i < \alpha + \beta \rangle$ where

$$Q_i = \begin{array}{l} R_i \text{ if } i < \alpha \\ S_j \text{ if } i = \alpha + j \end{array}$$

If $\mathbf{R} = (R_1, ..., R_m)$ and $\mathbf{S} = (S_1, ..., S_n)$ then $\mathbf{RS} = (R_1, ..., R_m, S_1, ..., S_n)$. Note that $\mathbf{RS}$ has the same range of definition as $\mathbf{R}$ and $\mathbf{S}$.

As we have pointed out in section 7.1, if $\mathbf{R} = \langle R_i : i < \alpha \rangle$ then $\mathbf{R}(A) =$
$= \langle A, \langle R_i(A) : i < \alpha \rangle \rangle$

Proposition 7.4.1: $\mathbf{RS}(A) = \mathbf{R}(A)\mathbf{S}(A)$ for all $A \in \mathbb{D}$.
The proof is omitted.

Corollary 7.4.2: If $\mathbf{R}(A) = A_1$ and $\mathbf{S}(A) = A_2$, then $\mathbf{RS}(A) = A_1 A_2$.

Proposition 7.4.3: If $A \in \mathbb{C}_\mathbf{RS}$ then there are $A_1 \in \mathbb{C}_\mathbf{R}$ and $A_2 \in \mathbb{C}_\mathbf{S}$ such that $A = A_1 A_2$.
Proof: According to the definition of $\mathbb{C}_\mathbf{RS}$ there is $C \in \mathbb{D}$ and $\varphi \in \mathrm{Bi}(C, A)$ such that

$A=\varphi[RS(C)]$. Since $RS(C)=R(C)S(C)$ it follows from lemma 2.8.1 that
$A=\varphi[R(C)]\varphi[S(C)]$. Let $\varphi[R(C)]=A_1$ and $\varphi[S(C)]=A_2$. Thus, $A=A_1A_2$ and $A_1 \in \mathbb{C}_R$ and $A_2 \in \mathbb{C}_S$. ♦

Proposition 7.4.4: For all $A \in \mathbb{C}_R$ and $B \in \mathbb{C}_S$, AR^SB iff $AB \in \mathbb{C}_{RS}$.
The proof is omitted.

Proposition 7.4.5: For all $A,B \in \mathbb{D}$
$$I_{RS}(A,B) = I_R(A,B) \cap I_S(A,B).$$
Proof: Suppose $R = \langle R_i : i < \alpha \rangle$ and $S = \langle S_i : i < \beta \rangle$. According to 7.1.1,
$$I_R(A,B) = \cap_{i<\alpha} I_{R_i}(A,B)$$
and
$$I_S(A,B) = \cap_{i<\beta} I_{S_i}(A,B).$$
$RS = \langle Q_i : i < \alpha+\beta \rangle$ where
$$Q_i = \begin{array}{l} R_i \text{ if } i < \alpha \\ \\ S_j \text{ if } i = \alpha+j. \end{array}$$
From 7.1.1 and the definition of RS follows
$$I_{RS}(A,B) = \cap_{i<\alpha+\beta} I_{Q_i}(A,B) = (\cap_{i<\alpha} I_{R_i}(A,B)) \cap (\cap_{i<\beta} I_{S_i}(A,B)) =$$
$$= I_R(A,B) \cap I_S(A,B). ♦$$

Corollary 7.4.6: $RsubRS$ and $SsubRS$.
Proof:
$$I_{RS}(A,B) = I_R(A,B) \cap I_S(A,B) \subseteq I_R(A,B).$$
Analogously,
$$I_{RS}(A,B) \subseteq I_S(A,B). ♦$$

Proposition 7.4.7: If $QsubR$ and $QsubS$, then $QsubRS$.
Proof:
$$I_Q(A,B) \supseteq I_R(A,B) \supseteq I_{RS}(A,B).$$
and
$$I_Q(A,B) \supseteq I_S(A,B) \supseteq I_{RS}(A,B). ♦$$

Proposition 7.4.8: If $Suncorr$**QR**, then $Suncorr$**Q** and $Suncorr$**R**. If $Scorr$**Q** or $Scorr$**R**, then $Scorr$**QR**.

Proof: From 7.4.6 follows **Q**sub**QR**. Thus, if $Suncorr$**QR** then according to 7.3.21 $Suncorr$**Q**. That $Suncorr$**R** is proved analogously. If $Scorr$**Q**, then $Scorr$**QR** according to 7.3.19. ◆

Corollary 7.4.9: If $\mathbb{D}_\mathbf{Q}$ is non-degenerate, then **Q**$corr$**QR**.

Proof: Since $\mathbb{D}_\mathbf{Q}$ is non-degenerate, **Q**$corr$**Q** according to 7.3.28. It then follows from 7.4.9 that **Q**$corr$**QR**. ◆

Proposition 7.4.10: For every $A_1 \in \mathbb{C}_\mathbf{Q}$ and $A_2 \in \mathbb{C}_\mathbf{R}$ such that $A_1 = A_2$,
$$(\mathbf{QR})^S[A_1 A_2] \subseteq \mathbf{Q}^S[A_1] \cap \mathbf{R}^S[A_2].$$
Proof: Suppose $A \in (\mathbf{QR})^S[A_1, A_2]$. Hence $A_1 A_2 (\mathbf{QR})^S A$ and $A = A_1 = A_2$. Then there is $B \in \mathbb{D}$ and $\varphi \in \mathbb{B}i(B, A)$ such that $A_1 A_2 = \varphi[\mathbf{QR}(B)]$ and $A = \varphi[S(B)]$. From this it follows that $A_1 = \varphi[\mathbf{Q}(B)]$ and $A_2 = \varphi[\mathbf{R}(B)]$ and, furthermore, $A_1 \mathbf{Q}^S A$ and $A_2 \mathbf{R}^S A$. Hence, $A \in \mathbf{Q}^S[A_1]$ and $A \in \mathbf{R}^S[A_2]$. ◆

Proposition 7.4.11: For every $A \in \mathbb{D}$,
$$(\mathbf{QR})^S[A] \subseteq \mathbf{Q}^S[A] \cap \mathbf{R}^S[A]$$
Proof: According to 7.4.10,
$$(\mathbf{QR})^S[\mathbf{Q}(A)\mathbf{R}(A)] \subseteq \mathbf{Q}^S[A] \cap \mathbf{R}^S[A].$$
Since $\mathbf{Q}(A)\mathbf{R}(A) = \mathbf{QR}(A)$ according to 7.4.1 the proposition follows. ◆

7.5   Significance

The study of relevance begun in section 4.9 is continued in this section. The investigation is centred around the notion of significance and transitions are our foremost formal device.

Let us say that **R** is *significant* for **S** with respect to **Q** if there are $X \in \mathbb{C}_\mathbf{Q}$, $Y \in \mathbb{C}_\mathbf{R}$ and $Z \in \mathbb{C}_\mathbf{S}$ such that $X\mathbf{Q}^\mathbf{R}Y$ and $X\mathbf{Q}^\mathbf{S}Z$ and it is not the case that $XY(\mathbf{QR})^S Z$. If **R** is not significant for **S** with respect to **Q** we say that **R** is *insignificant* for **S** with respect to **Q**. **R** is thus insignificant for **S** with respect to **Q** if for all $X \in \mathbb{C}_\mathbf{Q}$, $Y \in \mathbb{C}_\mathbf{R}$ and $Z \in \mathbb{C}_\mathbf{S}$, $X\mathbf{Q}^\mathbf{R}Y$ and $X\mathbf{Q}^\mathbf{S}Z$ implies $XY(\mathbf{QR})^S Z$.

Proposition 7.5.1: **R** is significant for **S** with respect to **Q** iff there are $X \in \mathbb{C}_\mathbf{Q}$ and $Y \in \mathbb{C}_\mathbf{R}$ such that $X\mathbf{Q}^\mathbf{R}Y$ and $(\mathbf{QR})^S[XY] \subset \mathbf{Q}^S[X]$.

Proof: (I) Suppose R is significant for S with respect to Q. Then there are $X \in \mathbb{C}_Q$, $Y \in \mathbb{C}_R$ and $Z \in \mathbb{C}_S$ such that XQ^RY and XQ^SZ and it is not the case that $XY(QR)^SZ$. According to 7.4.10, $(QR)^S[XY] \subseteq Q^S[X]$. It remains to prove that $(QR)^S[XY] \neq Q^S[X]$. This follows since $Z \in Q^S[X]$ but $Z \notin (QR)^S[XY]$.
(II) Suppose there are $X \in \mathbb{C}_Q$ and $Y \in \mathbb{C}_R$ such that XQ^RY and $(QR)^S[XY] \subset \subset Q^S[X]$. Then there is $Z \in Q^S[X]$ such that $Z \notin (QR)^S[XY]$. Thus XQ^SZ and $XY(QR)^SZ$. ♦

Proposition 7.5.2: R is significant for S with respect to Q iff there is a $A \in \mathbb{D}$ such that $(QR)^S[A] \subset Q^S[A]$.
Proof: (I) Suppose R is significant for S with respect to Q. According to 7.5.1 there are $X \in \mathbb{C}_Q$ and $Y \in \mathbb{C}_R$ such that XQ^RY and $(QR)^S[XY] \subset Q^S[X]$. Then there are $A \in \mathbb{D}$ and $\varphi \in \mathbb{B}i(A,X)$ such that $\varphi[Q(A)]=X$ and $\varphi[R(A)]=Y$. Thus, $XY = = \varphi[Q(A)]^\wedge \varphi[R(A)] = \varphi[QR(A)]$ according to 7.4.1 and therefore $(QR)^S[XY] = = (QR)^S[\varphi[QR(A)]] = \{ \varphi[Z] \mid Z \in (QR)^S[A] \}$ according to 7.2.5. By analogy we get $Q^S[X] = Q^S[\varphi[Q(A)]] = \{ \varphi[Z] \mid Z \in Q^S[A] \}$. Suppose now that $(QR)^S[A] = = Q^S[A]$. Then $(QR)^S[XY] = Q^S[X]$ which contradicts the assumption. This shows that $(QR)^S[A] \neq Q^S[A]$ and since $(QR)^S[A] \subseteq Q^S[A]$ according to 7.4.11 it follows that $(QR)^S[A] \subset Q^S[A]$.
(II) Suppose there is $A \in \mathbb{D}$ such that $(QR)^S[A] \subset Q^S[A]$. Since, according to 7.4.1, $QR(A)=Q(A)R(A)$, it follows that $(QR)^S[Q(A)R(A)] \subset Q^S[Q(A)]$. Note that $Q(A) \in \mathbb{C}_Q$ and $R(A) \in \mathbb{C}_R$ and $Q(A)Q^RR(A)$, which completes the proof.♦

From 7.5.2 it follows that R is not significant for S with respect to Q if for all $A \in \mathbb{D}$ it holds that $(QR)^S[A] = Q^S[A]$.

That R is significant for S with respect to Q we write as $RsignS \mid Q$. If R is insignificant for S with respect to Q we write $RinsignS \mid Q$.

We let RelS $\mathbb{D}$ denote the class of all relational systems with $\mathbb{D}$ as range of definition.

Proposition 7.5.3: If $SsubQ$ then for all $R \in$ RelS $\mathbb{D}$, $RinsignS \mid Q$.
Proof: Since $SsubQ$, $Q^S[A]=\{S(A)\}$ for all $A \in \mathbb{D}$ according to 7.2.8. Thus, $(QR)^S[A] = Q^S[A]$ for all $A \in \mathbb{D}$, which implies $RinsignS \mid Q$. ♦

Corollary 7.5.4: $RinsignS \mid S$.

Proposition 7.5.5: Suppose $RsubP$. Then $RsignS \mid Q$ implies $PsignS \mid Q$.
Proof: According to 7.5.2 there is $A \in \mathbb{D}$ such that $(QR)^S[A] \subset Q^S[A]$. $RsubP$ implies $QRsubQP$ and thus $(QP)^S[A] \subseteq (QR)^S[A]$. From this it follows that $(QP)^S[A] \subset Q^S[A]$ and, hence $PsignS \mid Q$. ◆

Corollary 7.5.6: Suppose $PsubR$. Then $RinsignS \mid Q$ implies $PinsignS \mid Q$.

Proposition 7.5.7: If $RsubQ$, then $RinsignS \mid Q$ for all $S \in RelS \; \mathbb{D}$.
Proof: $RsubQ$ implies $QparQR$, and thus $(QR)^S[A] = Q^S[A]$ and for all $A \in \mathbb{D}$. This shows that $RsignS \mid Q$. ◆

Corollary 7.5.8: If $RsubQ$, then $RinsignR \mid Q$.

Corollary 7.5.9: $RinsignS \mid R$.

Proposition 7.5.10: If not $RsubQ$, then $RsignR \mid Q$.
Proof: Since Q^R is not a function, it follows according to 7.2.8 that there is $A \in \mathbb{D}$ such that $Q^R[A] \supset \{R(A)\}$. According to 7.4.10 and 7.3.24 it follows that $(QR)^R[A] \subseteq Q^R[A] \cap R^R[A] = \{R(A)\}$, and thus $(QR)^R[A] = \{R(A)\}$. So $(QR)^R[A] \subset Q^R[A]$, and according to 7.5.2, $RsignR \mid Q$. ◆

Corollary 7.5.11: $RsubQ$ iff $RinsignR \mid Q$.
Proof: Use 7.5.10 and 7.5.8. ◆

Proposition 7.5.12: If $RsignS \mid Q$, then not $SsubQ$ but $ScorrQR$.
Proof: According to 7.5.2, there is $A \in \mathbb{D}$ such that $(QR)^S[A] \subset Q^S[A]$. Thus, according to 7.2.8, not $SsubQ$. According to 7.5.1, there are $X \in C_Q$ and $Y \in C_R$ such that XQ^RY and $(QR)^S[XY] \subset Q^S[X]$. From this it follows that $(QR)^S$ is not sweeping, and thus $ScorrQR$. ◆

The definitions of s-significance, c-significance and strong significance which were introduced in section 4.9 and there applied to relationals can be immediately generalized to systems of relationals. We use $Rs\text{-}signS \mid Q$ as an abbreviation for "R is s-significant for S with respect to Q" and analogously for the other significance and insignificance relations. Thus the following holds:

Rs-$sign$S $|$ Q iff SsubQR but not SsubQ.

Rc-$sign$S $|$ Q iff S$corr$QR but S$uncorr$Q.

Rs-$insign$S $|$ Q iff not SsubQR or SsubQ.

Rc-$insign$S $|$ Q iff S$uncorr$QR or S$corr$Q.

Proposition 7.5.13: If Rs-$sign$S $|$ Q then R$sign$S $|$ Q.
Proof: Suppose SsubQR but not SsubQ. Then according to 7.2.8 $(QR)^S[A] =$ $= \{S(A)\}$ for all $A \in \mathbb{D}$ and there is $B \in \mathbb{D}$ such that $Q^S[B] = \{S(B)\}$. Thus, $(QR)^S[B] \subset Q^S[B]$, and according to 7.5.2, R$sign$S $|$ Q. ◆

Proposition 7.5.14: If Rc-$sign$S $|$ Q then R$sign$S $|$ Q.
Proof: Suppose S$corr$QR but S$uncorr$Q. Then Q^S is sweeping but $(QR)^S$ is not sweeping. Thus, there are $X \in \mathbb{C}_Q$, $Y \in \mathbb{C}_R$ and $Z \in \mathbb{C}_S$ such that $X = Y = Z$ and $XY \in \mathbb{C}_{QR}$ but not $XY(QR)^S Z$. Since Q^S is sweeping, $XQ^S Z$. Therefore, according to the definition of significance, R$sign$S $|$ Q. ◆

Proposition 7.5.15: If R$sign$S $|$ Q and SsubQR then not SsubQ.
Proof: $(QR)^S[A] = \{S(A)\}$ for all $A \in \mathbb{D}$ and there is $B \in \mathbb{D}$ such that $Q^S[B] \supset$ $\supset (QR)^S[B]$. Thus, $Q^S[B] \supset \{S(B)\}$ which implies that Q^S is not a function. ◆

Corollary 7.5.16: If R$sign$S $|$ Q and SsubQR then Rs-$sign$S $|$ Q.

Proposition 7.5.17: If R$insign$S $|$ Q and S$uncorr$Q then S$uncorr$QR.
Proof: Suppose $XY \in \mathbb{C}_{QR}$ and $Z \in \mathbb{C}_S$ such that $X = Y = Z$. Then $XQ^R Y$ and, since S$uncorr$Q, $XQ^S Z$. R$insign$S $|$ Q implies $(QR)^S[XY] = Q^S[X]$ and thus $XY(QR)^S Z$. This shows that S$uncorr$Q. ◆

Corollary 7.5.18: If R$insign$S $|$ Q and S$uncorr$Q then Rc-$insign$S $|$ Q.

Proposition 7.5.19: If R$insign$S $|$ Q and SsubQR then SsubQ.
Proof: $(QR)^S[A] = \{S(A)\}$ and $(QR)^S[A] = Q^S[A]$ for all $A \in \mathbb{D}$. Thus, $Q^S[A] =$ $= \{S(A)\}$ for all $A \in \mathbb{D}$ and Q^S is a function. ◆

Proposition 7.5.20: If R$corr$S and Q$uncorr$S then R$sign$S $|$ Q.

Proof: According to 7.4.8 $\mathbf{QR}corr\mathbf{S}$. Then there are $\mathbf{XY} \in \mathbb{C}_{\mathbf{QR}}$ and $\mathbf{Z} \in \mathbb{C}_{\mathbf{S}}$ such that X=Y=Z but it is not the case that $\mathbf{XY(QR)^S Z}$. From $\mathbf{XY} \in \mathbb{C}_{\mathbf{QR}}$ and $\mathbf{Q}uncorr\mathbf{S}$ it follows that $\mathbf{XQ^R Y}$ and $\mathbf{XQ^S Z}$. This implies $\mathbf{R}sign\mathbf{S} \mid \mathbf{Q}$. (We could alternatively have used 7.5.17.) ♦

It seems reasonable to say that **R** is *significant* for **S**, which we abbreviate as $\mathbf{R}sign\mathbf{S}$, if there is a **Q** such that $\mathbf{R}sign\mathbf{S} \mid \mathbf{Q}$. $\mathbf{R}sign\mathbf{S}$ is not trivial since according to 7.5.4 and 7.5.9, $\mathbf{R}insign\mathbf{S} \mid \mathbf{S}$ and $\mathbf{R}insign\mathbf{S} \mid \mathbf{R}$.

Theorem 7.5.21: If $\mathbf{R}corr\mathbf{S}$ then $\mathbf{R}sign\mathbf{S}$.
Proof: We show that $\mathbf{R}sign\mathbf{S} \mid \mathbf{Q}$ where **Q** such that $\mathbb{C}_{\mathbf{Q}}$ is degenerate. Since $\mathbf{R}corr\mathbf{S}$ there are $\mathbf{X} \in \mathbb{C}_{\mathbf{R}}$ and $\mathbf{Y} \in \mathbb{C}_{\mathbf{S}}$ such that X=Y but not $\mathbf{XR^S Y}$. There is $A \in \mathbb{D}$ and a bijection φ such that $X = \varphi[R(A)]$. Let $Z = \varphi[Q(A)]$. Hence $\mathbf{ZQ^R X}$. But it is not the case that $\mathbf{ZX(QR)^S Y}$ since $(QR)^S[ZX] \subseteq Q^S[Z] \cap R^S[X]$ (according to 7.4.10) and $Y \notin R^S[X]$. From the fact that $\mathbb{C}_{\mathbf{Q}}$ is degenerate it follows that Q^S is sweeping according to 7.3.29 and 7.3.11. Thus $\mathbf{ZQ^S Y}$. Then it follows that $\mathbf{R}sign\mathbf{S} \mid \mathbf{Q}$.

At the end of section 1.6 the connection between irrelevance and invariance was described in preliminary fashions as follows: α is irrelevant for β if [the structure of] β is invariant under transformations that change only [the structure of] α. We can now compare this with what has been said about significance. For this purpose, let us say that **R** is irrelevant for **S** with respect to **Q** iff **S** is invariant under transformations that may change **R** but do not change **Q**, which means that if $\varphi \in \mathbb{I}_{\mathbf{Q}}(A,B)$ then $\varphi \in \mathbb{I}_{\mathbf{S}}(A,B)$, i.e. $\mathbf{S}sub\mathbf{Q}$. If $\mathbf{S}sub\mathbf{Q}$, then according to 7.5.3, $\mathbf{R}insign\mathbf{S} \mid \mathbf{Q}$. This is an argument for insignificance as an explication of irrelevance.

7.6 Stability and monotonicity of s-functions

In this section we will continue the study of stability for s-functions which we commenced in section 4.7. We start off with a series of definitions. To simplify notation we stipulate the following convention. If $\mathbf{X} = \langle X, \rho_i \rangle_{i<\alpha}$ and $\mathbf{Y} = \langle Y, \sigma_i \rangle_{i<\alpha}$ are similar, then $\mathbf{X} \subseteq \mathbf{Y}$ means that $X \subseteq Y$ and for all i, $i<\alpha$, $\rho_i \subseteq \sigma_i$.

Suppose f is an s-function on K into L. If, for all **A** and **B** in K,

$$B/A \subseteq A/B \implies f(B)/A \subseteq f(A)/B \qquad \text{then f is } strongly\ monotone\ increasing$$

B/A ⊆ A/B ⇒ f(B)/A ⊇ f(A)/B	then f is *strongly monotone decreasing*
B/A = A/B ⇒ f(B)/A = f(A)/B	then f is *strongly stable*
B ⊆ A ⇒ f(B) ⊆ f(A)	then f is *upward monotone increasing*
B ⊆ A & A/B ⊆ B ⇒ f(A)/B ⊆ f(B)	then f is *downward monotone increasing*
B ⊆ A ⇒ f(B) ⊇ f(A)/B	then f is *upward monotone decreasing*
B ⊆ A & A/B ⊆ B ⇒ f(A) ⊇ f(B)	then f is *downward monotone decreasing*
B = A/B ⇒ f(B) ⊆ f(A)	then f is *upward stable*
B = A/B ⇒ f(B) ⊇ f(A)/B	then f is *downward stable*
B = A/B ⇒ f(B) = f(A)/B	then f is *weakly stable.*

If f is strongly monotone increasing *or* strongly monotone decreasing, then f is *strongly monotone.*

If f is upward monotone increasing *and* downward monotone increasing, then f is *weakly monotone increasing.*

If f is upward monotone decreasing *and* downward monotone decreasing, then f is *weakly monotone decreasing.*

If f is upward monotone increasing *or* upward monotone decreasing, then f is *upward monotone.*

If f is downward monotone increasing *or* downward monotone decreasing, then f is *downward monotone.*

Note that f is upward monotone increasing *and* upward monotone decreasing iff for all A and B in K

$$B \subseteq A \Rightarrow f(B) = f(A)/B.$$

Analogously, f is downward monotone increasing *and* downward monotone decreasing iff for all A and B in K

$$B \subseteq A \& A/B \subseteq B \Rightarrow f(A)/B = f(B).$$

We use here the term "strong stability" for that condition which has before been called "stability", since it seems reasonable to talk about various conditions as "stability conditions".

Let us give a few examples of the relations between these different monotonicity and stability conditions.

Strongly monotone ⇒ strongly stable.
Strongly monotone increasing ⇒ weakly monotone increasing.
Weakly monotone increasing or weakly monotone decreasing ⇒ weakly stable.

Upward monotone increasing or downward monotone decreasing ⇒ upward stable.
Upward monotone decreasing or downward monotone increasing ⇒ downward stable.
Upward monotone increasing & upward monotone decreasing ⇒ weakly stable.
Downward monotone increasing & downward monotone decreasing ⇒ weakly stable.

The converse does not hold for any of the above implications. Note however that if K such that A and B in K and A∩B≠Ø implies A/B in K, then:
Weakly monotone increasing ⇒ strongly monotone increasing.
The following propositions are of a rather different character.

f strongly monotone ⇒ [A, A/B∈K ⇒ f(A/B)=f(A)/B] .
f upward monotone increasing ⇒ [A, A/B∈K ⇒ f(A/B) ⊆ f(A)/B] .
f downward monotone increasing ⇒ [A, A/B∈K ⇒ f(A)/B ⊆ f(A/B)] .
f weakly stable ⇒ [A, A/B∈K ⇒ f(A/B) = f(A)/B] .

In section 3.3 we discussed the method for constructing the weak order associated with a semiorder. Let ζ be a semiorder and w(ζ) the weak order associated with ζ. The operation w, which is a function on the set of semiorders into the set of weak orders, is domain-preserving and thus an s-function. From 2.5.2 it follows that w is downward stable. Let sw be the operation which when applied to a semiorder ζ gives the associated strict weak order as a result, and let ew applied to ζ give the associated equivalence relation as a result. From 2.5.2 it follows that sw is upward stable while ew is downward stable. (These are results we have already used in section 3.3.)

A relational R is in a rudimentary sense an s-function on $\mathbb{D}_R$ onto $\mathbb{E}_R$. This view of relationals presupposes that we regard sets as a kind of rudimentary structure. The structure $\langle X,\rho_1,...,\rho_k \rangle$ is of the type $\langle \upsilon_1,...,\upsilon_k \rangle$ if ρ_i is a υ_i-ary relation on X. Let us say that the set X is a structure of type $\langle \rangle$ (where $\langle \rangle$ is the empty sequence). R is thus strongly monotone increasing if for all A and B in $\mathbb{D}_R$,

$$A/B \subseteq B/A \Rightarrow R(A)/B \subseteq R(B)/A. \quad (*)$$

Note that A/B is the same as A∩B, so the antecedent in (*) is equivalent to A∩B = = B∩A, which holds generally. (*) thus reduces to

$$R(A)/B \subseteq R(B)/A .$$

Since this holds for all A and B in $\mathbb{D}_R$ it also holds that

$$R(B)/A \subseteq R(A)/B .$$

R is thus strongly monotone increasing iff for all A and B in $\mathbb{D}_R$

$$R(A)/B = R(B)/A ,$$

i.e. iff R is stable.

Analogously we can prove that for relationals, the properties of being strongly monotone decreasing and strong stability are both equivalent to stability.

R is upward monotone increasing iff

$$B \subseteq A \Rightarrow R(B) \subseteq R(A)$$

i.e. iff R is upward stable as it is defined in section 3.1. R is downward monotone increasing iff

$$B \subseteq A \Rightarrow R(A)/B \subseteq R(B)$$

i.e. iff R is downward stable as it is defined in section 3.1. It is also easy to see that for relationals, being downward monotone increasing and upward stability are each equivalent to upward stability (as defined in section 3.1). Analogously, being upward monotone decreasing and downward stability are for relationals each equivalent to downward stability (as it is defined in section 3.1). Furthermore, for relationals, weak stability, weakly monotone increasing and weakly monotone decreasing are each equivalent to weak stability as defined in section 3.1. Finally, for relationals both upward monotonicity and downward monotonicity are each equivalent to directed stability (defined in section 3.1).

All the monotonicity and stability properties defined above for relationals are therefore equivalent to stability, weak stability, upward stability, downward stability or directed stability already introduced in section 3.1.

Suppose that $\mathbb{D}_R = \mathbb{D}_S$. Suppose further that R is upward stable and that $\mathbb{F}_{RS}$ is upward monotone increasing. The following sequence of implications shows that S is upward stable.

$$B \subseteq A \Rightarrow R(B) \subseteq R(A) \Rightarrow \mathbb{F}_{RS}(R(B)) \subseteq \mathbb{F}_{RS}(R(A)) \Rightarrow S(B) \subseteq S(A) .$$

Let us mention a few other results on the interplay between different monotonicity and stability properties for R,S and $\mathbb{F}_{RS}$. We presuppose here that $\mathbb{D}_R = \mathbb{D}_S$.

R weakly stable & $\mathbb{F}_{RS}$ upward stable $\Rightarrow$ S upward stable.

S upward stable $\Rightarrow$ $\mathbb{F}_{RS}$ upward monotone increasing & downward monotone decreasing & upward stable.

R downward stable & $\mathbb{F}_{RS}$ downward monotone increasing $\Rightarrow$ S downward stable.

R weakly stable & $\mathbb{F}_{RS}$ weakly stable $\Rightarrow$ S weakly stable.

R weakly stable & [$\mathbb{F}_{RS}$ weakly monotone increasing or weakly monotone decreasing] $\Rightarrow$ S weakly stable.

R stable & [$\mathbb{F}_{RS}$ strongly monotone increasing or strongly monotone decreasing or strongly stable] $\Rightarrow$ S stable.

CHAPTER 8

THE STRUCTURE OF SUBORDINATION

8.0 Introduction

Subordination, *sub,* is a relation between systems of relationals whose formal properties will be more closely examined in this chapter. We will see that it generates a lattice in a natural way, and one of our main tasks is to find out what kind of lattice. This is done in sections 8.1 and 8.2, where we introduce two new concepts, the partial order which *sub* generates and the notion of the rank of a relational system. In section 8.3 we will devote some effort to investigating what role the pair *corr* and *uncorr*, i.e. the relations *correlation* and *uncorrelation*, play in the lattice generated by *sub*. A generalization of the notion of a rank is studied in section 8.4. Already in chapter 1 I said that *sub* expresses the idea of equality preservation applied to structures. In section 8.5 this suggestion will be made more precise and we will see that *uncorr* expresses the idea of realizability of structures. It is tempting to regard *sub, corr* and *uncorr* themselves as relationals, but I shall not pursue this line of thought here.

8.1 Subalternation and rank

As was already mentioned in section 4.2 subordination, i.e. *sub*, is a preorder. There is a standard method for generating a partial order from a preorder. We shall now apply this method to *sub*.

Parity is defined by **R***par***S** iff **R***sub***S** and **S***sub***R**. Since parity is an equivalence relation we can construct the class of equivalence classes modulo parity. Let **R̲** be the equivalence class generated by **R**, and **S̲** the equivalence class generated by S. Define *subalternation* in the following way:

S *is subaltern to* **R** iff S*sub*R

It is well-known that the ordering relation defined from a preorder in this way is always a partial order (see for example Birkhoff,1967, p. 21). Subalternation is thus a partial order, and we denote it *sub* (compare subordination which is denoted *sub*). Note that

$$R = S \text{ iff } R \text{ } parS.$$

Note that S*sub*R implies $\mathbb{D}_R = \mathbb{D}_S$.

In chapter 7 we introduced RelS $\mathbb{D}$ to denote the set of relational systems with $\mathbb{D}$ as range of definition. We supplement this terminology with two new symbols. Let Rel $\mathbb{D}$ be the set of relationals with $\mathbb{D}$ as range of definition and let RelS $\mathbb{D}$ be the set of equivalence classes modulo parity generated from RelS $\mathbb{D}$. Since $\langle$RelS $\mathbb{D}$, *sub*$\rangle$ is a preorder it follows that $\langle$RelS $\mathbb{D}$, *sub*$\rangle$ is a partial order.

There is a natural representation of RelS $\mathbb{D}$, a representation that can be rather useful. To study this we introduce the notion of rank. **R** and S are on a par iff $\mathbb{D}_R = \mathbb{D}_S$ and for all A,B$\in \mathbb{D}$ (where $\mathbb{D} = \mathbb{D}_R = \mathbb{D}_S$)

$$\mathbb{I}(R(A),R(B)) = \mathbb{I}(S(A),S(B)).$$

If **R** and S are on a par none is subordinate to the other, and using a military metaphor, it is natural to say that **R** and S have the same rank. What is the thing **R** and S have the same of (i.e. have in common) when they are on a par? The answer is rather obvious: the set of isomorphisms from the extension of A to the extension of B for all A,B$\in \mathbb{D}$. We have previously used $\mathbb{I}_R(A,B)$ just as an abbreviation of $\mathbb{I}(R(A),R(B))$, but we shall now let $\mathbb{I}_R$ "live its own life", i.e. regard $\mathbb{I}_R$ as a function with $\mathbb{D} \times \mathbb{D}$ as domain. Let us define the rank of a relational system in the following way: The *rank* of **R** is the function $\mathbb{I}_R$ on $\mathbb{D}_R \times \mathbb{D}_R$ such that

$$\mathbb{I}_R(A,B) = \mathbb{I}(R(A),R(B)).$$

R and S are obviously on a par iff $\mathbb{I}_R = \mathbb{I}_S$. Note that $\mathbb{I}_R(A,B) \subseteq \mathbb{B}i(A,B)$.

A rank of a relational system is thus a function which takes ordered pairs of sets as arguments and as values sets of bijections between the components of the arguments. The set of ranks of the relational systems with $\mathbb{D}$ as range of definition will be denoted Rank $\mathbb{D}$. Hence,

$$\text{Rank } \mathbb{D} = \{\mathbb{I}_R \mid R \in \text{RelS } \mathbb{D}\}.$$

It is convenient for our purpose to regard Rank $\mathbb{D}$ as a subset of a wider class of functions with $\mathbb{D}$ as domain. Let us say that a *pseudo-rank defined for* $\mathbb{D}$ is a function F with $\mathbb{D} \times \mathbb{D}$ as domain such that

$$F(A,B) \subseteq \mathbb{B}i(A,B).$$

We denote the set of pseudo-ranks defined for $\mathbb{D}$ by Prank $\mathbb{D}$. If $F \in$ Prank $\mathbb{D}$ then we also say that F is a pseudo-rank *over* $\mathbb{D}$.

Equality,$=$, for pseudo-ranks is used in a straightforward way,
$$F=G \text{ iff } F(A,B) = G(A,B) \text{ for all } A,B \in \mathbb{D}.$$

If f and g are real-valued functions we can define "pointwise" addition and multiplication in the following way:
$$(f+g)(x) = f(x)+g(x)$$
$$(f \cdot g)(x) = f(x) \cdot g(x)$$

We can now define pointwise union and intersection for pseudo-ranks in an analogous way. Let $F, G \in$ Prank$(\mathbb{D})$. Define $\cup$ and $\cap$ in the following way: For all $A,B \in \mathbb{D}$
$$(F \cup G)(A,B) = F(A,B) \cup G(A,B)$$
$$(F \cap G)(A,B) = F(A,B) \cap G(A,B)$$

We define pointwise complement , $\sim$, as the complement of $F(A,B)$ with respect to $\mathbb{B}i(A,B)$: For all $A,B \in \mathbb{D}$
$$(\sim F)(A,B) = \mathbb{B}i(A,B) \setminus F(A,B)$$

In an analogous way we define pointwise inclusion as follows: For all $A,B \in \mathbb{D}$
$$F \subseteq G \text{ iff } F(A,B) \subseteq G(A,B)$$

The "pointwise" operations and relations defined above behave in an analogous way to the corresponding "ordinary" operation. It holds for example that
$$F \subseteq G \text{ iff } F \cap G = F.$$

Let $\mathbb{B}i$ and $\varnothing$ be the pseudo-ranks defined in the following obvious way: For all $A,B \in \mathbb{D}$
$$(\mathbb{B}i)(A,B) = \mathbb{B}i(A,B)$$
$$\varnothing(A,B) = \varnothing.$$

The following theorem, which characterizes pseudo-ranks, comes as no surprise.

Theorem 8.1.1: $\langle$ Prank $\mathbb{D}, \cup, \cap, \sim, \varnothing, \mathbb{B}i \rangle$ is a Boolean algebra and $\subseteq$ the corresponding partial order.

The proof, which is straightforward, is omitted.

Rank $\mathbb{D}$ is of course a subset of Prank $\mathbb{D}$, so the above pointwise operations and relations can of course be applied to ranks. If $R,S \in$ RelS $\mathbb{D}$ then for all $A,B \in \mathbb{D}$
$$(\mathbb{I}_R \cap \mathbb{I}_S)(A,B) = \mathbb{I}_R(A,B) \cap \mathbb{I}_S(A,B)$$
$$(\mathbb{I}_R \cup \mathbb{I}_S)(A,B) = \mathbb{I}_R(A,B) \cup \mathbb{I}_S(A,B)$$

$$(\sim\mathbb{I_R})(A,B) = \mathbb{B}i(A,B) \setminus \mathbb{I_R}(A,B)$$

$\mathbb{I_R} \cap \mathbb{I_S}$, $\mathbb{I_R} \cup \mathbb{I_S}$ and $\sim\mathbb{I_R}$ are of course pseudo-ranks over $\mathbb{D}$. But are they also ranks? $\mathbb{I_R} \cap \mathbb{I_S}$ is always a rank. This follows since for all $A,B \in \mathbb{D}$,

$$\mathbb{I_{RS}}(A,B) = \mathbb{I_R}(A,B) \cap \mathbb{I_S}(A,B) = (\mathbb{I_R} \cap \mathbb{I_S})(A,B)$$

which implies that

$$\mathbb{I_{RS}} = \mathbb{I_R} \cap \mathbb{I_S} .$$

$\mathbb{I_R} \cap \mathbb{I_S}$ is thus always a rank over $\mathbb{D}$ if $\mathbf{R},\mathbf{S} \in \mathrm{RelS}\ \mathbb{D}$, but the situation is different with $\mathbb{I_R} \cup \mathbb{I_S}$ and $\sim\mathbb{I_R}$. If $\mathbb{I_R} \in \mathrm{Rank}\ \mathbb{D}$, then it is not certain that $\sim\mathbb{I_R} \in \mathrm{Rank}\ \mathbb{D}$. The reason is that $(\sim\mathbb{I_R})(A,B)$, which is $\mathbb{B}i(A,B) \setminus \mathbb{I_R}(A,B)$, need not be the set of isomorphisms from some structure $\mathbf{A}$ to some structure $\mathbf{B}$, which is necessary for $\sim\mathbb{I_R}$ to be the rank of a relational system. The situation is analogous for $\mathbb{I_R} \cup \mathbb{I_S}$. $\mathbb{I_R}(A,B) \cup \mathbb{I_S}(A,B)$ is not generally the set of isomorphisms from some structure $\mathbf{A}$ to some structure $\mathbf{B}$, so it does not hold generally that $\mathbb{I_R} \cup \mathbb{I_S}$ is the rank of a relational system. We shall return to this in section 8.2.

$\mathbb{B}i$ is of course a member of Rank $\mathbb{D}$, since $\mathbb{B}i$ is the rank of for example the n-ary universal relational V defined by $V(A)=A^n$. On the other hand, $\emptyset$ does not belong to Rank $\mathbb{D}$. The reason is that for every relational system $\mathbf{R}$, the identity function ι_A on A is an isomorphism from $\mathbf{R}(A)$ to $\mathbf{R}(A)$. Hence, $\mathbb{I_R}(A,A) \neq \emptyset$. $\langle$ Rank $\mathbb{D}, \cap, \sim \rangle$ is as we have seen not generally a Boolean algebra. We shall return to the question what formal properties it does have in section 8.2.

Since $\langle$ Prank $\mathbb{D}, \supseteq \rangle$ is a partial order the same holds for $\langle$ Rank $\mathbb{D}, \supseteq' \rangle$ where $\supseteq'$ is $\supseteq$ restricted to Rank $\mathbb{D}$. For simplicity we denote in many cases $\supseteq'$ with just $\supseteq$. This is a convention we have already tacitly supposed for $\cap, \cup$ and $\sim$. Note that $\mathbb{I_S} \supseteq \mathbb{I_R}$ iff for all $A,B \in \mathbb{D}$, $\mathbb{I_S}(A,B) \supseteq \mathbb{I_R}(A,B)$. Hence,

$$\mathbb{I_S} \supseteq \mathbb{I_R} \quad \text{iff} \quad \mathbf{S}\mathit{sub}\mathbf{R} .$$

The reason for introducing ranks is that they function as a natural representation of relational systems. Let $\mathfrak{S}$ be the function which maps a relational system on its rank. Hence, $\mathfrak{S}$ is the function on RelS $\mathbb{D}$ into Rank $\mathbb{D}$ such that

$$\mathfrak{S}(\mathbf{R}) = \mathbb{I_R} .$$

Let $\underline{\mathfrak{S}}$ be the function on **RelS** $\mathbb{D}$ into Rank $\mathbb{D}$ such that

$$\underline{\mathfrak{S}}(\mathbf{R}) = \mathbb{I_R} .$$

We formulate the fundamental property of $\mathfrak{S}$ and $\underline{\mathfrak{S}}$ as a theorem.

Theorem 8.1.2: $\mathfrak{S}$ is a homomorphism from $\langle$ RelS $\mathbb{D}, \mathit{sub} \rangle$ to $\langle$ Rank $\mathbb{D}, \supseteq \rangle$

and $\mathfrak{S}$ an isomorphism from $\langle \underline{\text{RelS}}\ \mathbb{D}, \underline{sub} \rangle$ to $\langle \text{Rank}\ \mathbb{D}, \supseteq \rangle$.

Proof: $\mathfrak{S}$ is obviously onto Rank $\mathbb{D}$. That $\mathfrak{S}$ is a homomorphism then follows from the equivalence

$$S \underline{sub} R \quad \text{iff} \quad \mathbb{I}_S \supseteq \mathbb{I}_R .$$

Since $S \underline{par} R$ iff $\mathbb{I}_S = \mathbb{I}_R$, it follows that $\underline{S} = \underline{R}$ iff $\mathbb{I}_S = \mathbb{I}_R$. Hence, $\mathfrak{S}$ is bijective. That $\mathfrak{S}$ is an isomorphism then follows from the following series of equivalences:

$$\underline{S \underline{sub} R} \quad \text{iff} \quad S \underline{sub} R \quad \text{iff} \quad \mathbb{I}_S \supseteq \mathbb{I}_R . \quad \blacklozenge$$

Since $\langle \underline{\text{RelS}}\ \mathbb{D}, \underline{sub} \rangle$ and $\langle \text{Rank}\ \mathbb{D}, \supseteq \rangle$ are isomorphic we can study the set of ranks instead of the set of relational systems modulo parity. There must therefore exist an operation on $\underline{\text{RelS}}\ \mathbb{D}$ corresponding to $\cup$. We shall study it in the next section, where we examine the formal properties of subalternation.

8.2 The lattice of subalternation

In this section we study the formal properties of $\langle \underline{\text{RelS}}\ \mathbb{D}, \underline{sub} \rangle$. As was pointed out in the preceding section, $\langle \underline{\text{RelS}}\ \mathbb{D}, \underline{sub} \rangle$ is isomorphic to $\langle \text{Rank}\mathbb{D}, \supseteq \rangle$ and we can therfore alternatively study $\langle \text{Rank}\ \mathbb{D}, \supseteq \rangle$, and we shall frequently do this. It is important in this context that $\supseteq$ is, to be exact, the restriction of $\supseteq$ (as a relation on Prank $\mathbb{D}$) to Rank $\mathbb{D}$. To emphasize this we use here a special symbol for $\supseteq$ restricted to $\mathbb{D}$, viz, $\geq$. Thus,

$$\mathbb{I}_S \geq \mathbb{I}_R \quad \text{iff} \quad \mathbb{I}_S \supseteq \mathbb{I}_R \quad \text{iff} \quad \underline{S \underline{sub} R} \quad \text{iff} \quad S \underline{sub} R.$$

As will be clear later on, it is convenient to take the converse of $\geq$, which we denote by $\leq$, and thus study the partial order $\langle \text{Rank}\ \mathbb{D}, \leq \rangle$. Note that

$$\mathbb{I}_R \leq \mathbb{I}_S \quad \text{iff} \quad S \underline{sub} R \quad \text{iff} \quad \mathbb{I}_R(A,B) \subseteq \mathbb{I}_S(A,B) \quad \text{for all } A,B \in \mathbb{D}$$

$$\mathbb{I}_R = \mathbb{I}_S \quad \text{iff} \quad R \underline{par} S \quad \text{iff} \quad \mathbb{I}_R(A,B) = \mathbb{I}_S(A,B) \quad \text{for all } A,B \in \mathbb{D}$$

$\langle \text{Prank}\ \mathbb{D}, \cup, \cap, \sim, \varnothing, \text{Bi} \rangle$ is a Boolean algebra. Hence, each pair of pseudo-ranks $\{F,G\}$ over $\mathbb{D}$ has a least upper bound and a greatest lower bound, in both cases with respect to the partial order $\subseteq$. To be more exact, it holds that

$$\text{lub}\ \{F,G\} = F \cup G$$

$$\text{glb}\ \{F,G\} = F \cap G,$$

where lub and glb is with respect to $\subseteq$. Even each pair of ranks over $\mathbb{D}$ has a least upper bound and a greatest lower bound with respect to $\leq$. glb $\{\mathbb{I}_R, \mathbb{I}_S\}$ is the same with respect to both $\subseteq$ and $\leq$, but for lub$\{\mathbb{I}_R, \mathbb{I}_S\}$ the situation is more complicated. We shall look at this in detail. First however some preliminaries.

In section 8.1 we defined $\cup$ and $\cap$ for pseudo-ranks. We can in a straight-forward way generalize these operations to collections of pseudo-ranks. Suppose J is a collection of pseudo-ranks over $\mathbb{D}$. Then $\cup J$ and $\cap J$ are defined as follows. For all $A, B \in \mathbb{D}$,

$$[\cup J](A,B) = \cup \{F(A,B) \mid F \in J\}$$
$$[\cap J](A,B) = \cap \{F(A,B) \mid F \in J\}.$$

$\cup J$ and $\cap J$ are of course pseudo-ranks over $\mathbb{D}$.

Lemma 8.2.1: Suppose $J = \{F_i \mid i \in I\}$ where $F_i \in \text{Prank } \mathbb{D}$ for all $i \in I$. Then
(i) $F_i \supseteq G$ for all $i \in I$ implies $\cap J \supseteq G$.
(ii) $G \in J$ implies $G \supseteq \cap J$.
Proof: According to the definition of $\cap J$

$$[\cap J](A,B) = \cap \{F_i(A,B) \mid i \in I\}.$$

If $F_i \supseteq G$ then $F_i(A,B) \supseteq G(A,B)$ and since this holds for all $i \in I$ it follows that

$$[\cap J](A,B) \supseteq G(A,B).$$

This proves (i).

If $G \in J$ then $G = F_j$ for some $j \in I$. Hence, $G(A,B) \supseteq [\cap J](A,B)$ for all $A, B \in \mathbb{D}$, which proves (ii). $\blacklozenge$

Lemma 8.2.2: If $\mathbf{R} = \langle R_i : i < \alpha \rangle$ then

$$\mathbb{I}_{\mathbf{R}} = \cap_{i<\alpha} \mathbb{I}_{R_i}$$

Proof: According to 7.1.1

$$\mathbb{I}_{\mathbf{R}}(A,B) = \cap_{i<\alpha} \mathbb{I}_{R_i}(A,B)$$

Since

$$\cap_{i<\alpha} \mathbb{I}_{R_i}(A,B) = [\cap_{i<\alpha} \mathbb{I}_{R_i}](A,B)$$

the lemma follows. $\blacklozenge$

Corollary 8.2.3: If $\mathbf{R} = \langle R_i : i < \alpha \rangle$ then for all $i < \alpha$

$$\mathbb{I}_{\mathbf{R}} \leq \mathbb{I}_{R_i}$$

and

$$\mathbb{I}_{\mathbf{R}}(A,B) \subseteq \mathbb{I}_{R_i}(A,B)$$

whenever $A, B \in \mathbb{D}$.

Lemma 8.2.4: If $J \subseteq \text{Prank } \mathbb{D}$, then $\cap J = \text{glb} J$ (with respect to $\subseteq$).

Proof: Suppose $F \in J$. According to 8.2.1, $\cap J \subseteq F$. Hence, $\cap J$ is a lower bound for J. Suppose G is a lower bound for J. Then $G \subseteq F$ for all $F \in J$. According to 8.2.1, $\cap J \supseteq G$. $\cap J$ is thus the greatest lower bound for J. ◆

From 8.2.4, 8.1.1 and the fact that $\cap J$ is a pseudorank follows:

Theorem 8.2.5: $\langle \text{Prank } \mathbb{D}, \subseteq \rangle$ is a complete Boolean algebra.

Lemma 8.2.6: If $J \subseteq \text{Rank } \mathbb{D}$, then $\cap J \in \text{Rank } \mathbb{D}$.
Proof: Rel $\mathbb{D}$ is the set of relationals with $\mathbb{D}$ as range of definition. Let
$$\Sigma = \{Q \in \text{Rel } \mathbb{D} \mid \exists\, \mathbf{R} = \langle R_i : i < \alpha \rangle \in \text{RelS } \mathbb{D}: \exists\, j < \alpha: \mathbb{I}_{\mathbf{R}} \in J \,\&\, Q = R_j\}$$
Let $\mathfrak{R}$ be a wellordering of Σ and suppose that $\mathfrak{R} = \langle Q_i : i < \beta \rangle$. Hence, $\mathfrak{R} \in \text{RelS } \mathbb{D}$ and $\mathbb{I}_{\mathfrak{R}} \in \text{Rank } \mathbb{D}$. According to 8.2.2,
$$\mathbb{I}_{\mathfrak{R}} = \cap_{i < \beta}\, \mathbb{I}_{Q_i}.$$
If $\mathbb{I}_S \in J$ then every relational which is a term in S is also a term in $\mathfrak{R}$, which implies that $\mathbb{I}_S \supseteq \mathbb{I}_{\mathfrak{R}}$. From 8.2.1 then follows $\cap J \supseteq \mathbb{I}_{\mathfrak{R}}$. To prove that $\cap J \subseteq \mathbb{I}_{\mathfrak{R}}$, note that for every term Q_i in $\mathfrak{R}$ there is $R_i \in J$ such that Q_i is a term in R_i. Hence, according to 8.2.3 $\mathbb{I}_{R_i}(A,B) \subseteq \mathbb{I}_{Q_i}(A,B)$ for all $i < \beta$ and all $A,B \in \mathbb{D}$. Hence,
$$\cap_{i < \beta}\, \mathbb{I}_{R_i}(A,B) \subseteq \cap_{i < \beta}\, \mathbb{I}_{Q_i}(A,B) = \mathbb{I}_{\mathfrak{R}}(A,B)$$
and since $R_i \in J$ for all $i < \beta$
$$\cap_{i < \beta}\, \mathbb{I}_{R_i}(A,B) \supseteq (\cap J)(A,B).$$
These equations give
$$(\cap J)(A,B) \subseteq \mathbb{I}_{\mathfrak{R}}(A,B),$$
which holds for all $A,B \in \mathbb{D}$. From this follows $\cap J \subseteq \mathbb{I}_{\mathfrak{R}}$. We have thus proved that $\cap J = \mathbb{I}_{\mathfrak{R}}$ and since $\mathbb{I}_{\mathfrak{R}} \in \text{Rank } \mathbb{D}$ it follows that $\cap J \in \text{Rank } \mathbb{D}$. ◆

8.2.6 shows that Rank $\mathbb{D}$ is a closure system (see section 2.11) in the complete lattice $\langle \text{Prank } \mathbb{D}, \subseteq \rangle$. From well-known results in lattice theory (see section 2.11) follows then immediately the following theorem.

Theorem 8.2.7: $\langle \text{Rank } \mathbb{D}, \leq \rangle$ is a complete lattice.

Suppose $J \subseteq \text{Rank } \mathbb{D}$. Then the greatest lower bound of J taken in Rank $\mathbb{D}$, i.e.

with respect to $\leq$, is $\cap J$ and thus the same as the greatest lower bound of J taken in Prank $\mathbb{D}$, i.e. with respect to $\subseteq$. However, the least upper bound of J taken in Rank $\mathbb{D}$ is not necessarily the same as the upper bound of J taken in Prank $\mathbb{D}$. lub J with respect to $\leq$ is the greatest lower bound of all upper bounds of J in Rank $\mathbb{D}$ (see section 2.11). If $J \subseteq \text{Rank } \mathbb{D}$ we denote the set of all upper bounds for J with respect to $\leq$ with $U(J)$. Hence,

$$U(J) = \{F \in \text{Rank } \mathbb{D} \mid F \geq G \text{ for all } G \in J\}.$$

Thus, $\text{lub} J = \cap U(J)$, where lub are with respect to $\leq$.

Let us simplify $U(\{F_1, F_2, ..., F_n\})$ to $U(F_1, F_2, ..., F_n)$. From the above discussion follows that

$$U(\mathbb{I}_R, \mathbb{I}_S) = \{\mathbb{I}_Q \mid Q \in \text{RelS } \mathbb{D} \ \& \ \mathbb{I}_Q \geq \mathbb{I}_R \ \& \ \mathbb{I}_Q \geq \mathbb{I}_S\}$$

and

$$\text{lub}\{\mathbb{I}_R, \mathbb{I}_S\} = \cap\{\mathbb{I}_Q \mid Q \in \text{RelS } \mathbb{D} \ \& \ \mathbb{I}_Q \geq \mathbb{I}_R \ \& \ \mathbb{I}_Q \geq \mathbb{I}_S\}.$$

Note that $\text{glb}\{\mathbb{I}_R, \mathbb{I}_S\} = (\mathbb{I}_R \cap \mathbb{I}_S) = \mathbb{I}_{RS}$.

Since $\langle \text{Rank } \mathbb{D}, \leq \rangle$ is a lattice we can define two operations $\wedge$ and $\vee$ on Rank $\mathbb{D}$ in the following way:

$$F \wedge G = \text{glb } \{F, G\}$$
$$F \vee G = \text{lub } \{F, G\}$$

where glb and lub are taken with respect to $\leq$. As we have seen, $\wedge$ is the same operation as $\cap$. However, $\vee$ is not the necessarily the same as $\cup$. $\text{lub}\{F, G\} = \cap U\{F, G\}$ which need not be the same as $F \cup G$. Note that

$$F \cup G \subseteq F \vee G.$$

This holds since $F \vee G$ is an upper bound for $\{F, G\}$ with respect to $\leq$ and therefore even with respect to $\subseteq$, while $F \cup G$ is a least upper bound with respect to $\subseteq$. ($\leq$ is a subset of $\subseteq$.)

Because of the isomorphism between ranks and the set of equivalence classes modulo parity the structure of subalternation is that of a complete lattice. Let us look at this in more detail. In section 7.3 we introduced sup as the converse of sub. Let us use $\underline{sup}$ as the converse of $\underline{sub}$. Then $\langle \underline{\text{RelS}} \mathbb{D}, \underline{sup} \rangle$ is isomorphic to $\langle \text{Rank } \mathbb{D}, \leq \rangle$ and thus a complete lattice. Let us use $\underline{\wedge}$ and $\underline{\vee}$ as the correspondence on $\underline{\text{RelS}} \mathbb{D}$ to $\wedge$ and $\vee$ respectively. Hence, $R \underline{\wedge} S = \text{glb}\{R, S\}$, $R \underline{\vee} S = \text{lub}\{R, S\}$ where glb and lub are with respect to $\underline{sup}$. Note that

$$R \underline{\wedge} S = \underline{RS} \qquad \& \qquad R \underline{\vee} S = \underline{\mathfrak{R}},$$

where $\mathfrak{R}$ contains every relational with $\mathbb{D}$ as range of definition which is subordinate to both $\underline{R}$ and $\underline{S}$.

8.3 Correlation and collaterality

In section 7.3 it was proved that *par* is a congruence relation for *corr* and *uncorr*. We can therefore define relations *corr* and *uncorr* for equivalent classes of relational systems with respect to parity according to the rules

$$\underline{R} \; \underline{uncorr} \; \underline{S} \quad \text{iff} \quad R \; uncorr \; S$$
$$\underline{R} \; \underline{corr} \; \underline{S} \quad \text{iff} \quad R \; corr \; S$$

Consequently, we can also define relations corresponding to *uncorr* and *corr* for ranks. Let us use $\propto$ for the relation corresponding to *uncorr* and ∞ for the relation corresponding to *corr*. We introduce the following definitions:

$$\mathbb{I}_S \propto \mathbb{I}_R \quad \text{iff} \quad S \; uncorr \; R$$
$$\mathbb{I}_S \infty \mathbb{I}_R \quad \text{iff} \quad S \; corr \; R.$$

These definitions might be said to be implicit, and it would be interesting to have more explicit or straightforward definitions. The following lemma shows how this can be done.

Lemma 8.3.1: *SuncorrR* iff for all $A,B \in D$, $\varphi \in Bi(B,A)$ implies that there is $C \in D$ and $\omega \in Bi(C,A)$ such that

$$\omega \in \mathbb{I}_R(C,A) \; \text{and} \; \varphi^{-1} o \, \omega \in \mathbb{I}_S(C,B).$$

Proof: (I) Suppose *SuncorrR* and $A,B \in D$ and $\varphi \in Bi(B,A)$. Thus, $R(A) \in C_R$, $\varphi[S(B)] \in C_S$ and the domain of $\varphi[S(B)]$ is A. According to 7.3.3 there is $C \in D$ and $\omega \in Bi(C,A)$ such that

$$R(A) = \omega[R(C)] \; \text{and} \; \varphi[S(B)] = \omega[S(C)].$$

Thus $\omega \in \mathbb{I}_R(C,A)$ and $\varphi^{-1} o \omega \in \mathbb{I}_S(C,B)$.

(II) Suppose the second part of the equivalence in the lemma. Suppose further that $X \in C_R$, $Y \in C_S$ and $X=Y$. Then there are $A,B \in D$, $\varphi \in Bi(A,X)$ and $\psi \in Bi(B,Y)$ such that

$$X = \varphi[R(A)] \; \text{and} \; Y = \psi[S(B)] \, .$$

Since $X=Y$, it holds that $\varphi^{-1} o \psi \in Bi(B,A)$. According to the assumption then it follows that there are $C \in D$ and $\omega \in Bi(C,A)$ such that there are $C \in D$ and $\omega \in Bi(C,A)$ such that

$$\omega \in \mathbb{I}_R(C,A) \; \text{and} \; (\varphi^{-1} o \psi)^{-1} \in \mathbb{I}_S(C,B) \, .$$

Hence $R(A) = \omega[R(C)]$ and $S(B) = (\psi^{-1} o \varphi o \omega)[S(C)]$. Therefore

$$X = (\varphi o \psi)[R(C)] \; \text{and} \; Y = (\varphi o \omega)[S(C)] \, .$$

From 7.3.3 it follows that $Suncorr\mathbf{R}$. ♦

From 8.3.1 it follows that we can define $\propto$ in the following way: For all $F,G \in Rank\mathbb{D}$, $F \propto G$ iff

for all $A,B \in \mathbb{D}$, if $\varphi \in \mathbb{B}i(B,A)$ then there are $C \in \mathbb{D}$ and $\omega \in \mathbb{B}i(C,A)$

such that $\omega \in F(C,A)$ and $\varphi^{-1} \circ \omega \in G(C,B)$. $\qquad$ (8.3.2)

Obviously,

$$F \infty G \quad \text{iff} \quad \text{not } F \propto G.$$

8.3.2 is of course meaningful even if F and G are pseudo-ranks and not ranks. We can therefore generalize the notion of $\propto$ to Prank $\mathbb{D}$ by saying that for all $F,G \in$ Prank $\mathbb{D}$, $F \propto G$ iff 8.3.2 holds.

It seems reasonable that "the more distant" $\mathbb{I}_{\mathbf{R}}$ and $\mathbb{I}_{\mathbf{S}}$ are from each other in the lattice Rank$\mathbb{D}$, then the weaker is the dependence between $\mathbf{R}$ and S. However, what is meant by the distance between elements in a lattice is not quite clear. We shall here only use the idea that if the least upper bound of two elements x and y is the greatest element in the lattice, then x and y are not close to each other. Let us therefore introduce the following notion. The relational systems $\mathbf{R}$ and S with the same range of definition $\mathbb{D}$ are said to be *collateral* if $\mathbb{I}_{\mathbf{R}} \vee \mathbb{I}_{\mathbf{S}} = \mathbb{B}i$, where $\mathbb{I}_{\mathbf{R}} \vee \mathbb{I}_{\mathbf{S}}$ is the least upper bound of $\mathbb{I}_{\mathbf{R}}$ and $\mathbb{I}_{\mathbf{S}}$ in Rank $\mathbb{D}$. That $\mathbf{R}$ and S are *collateral* is abbreviated $\mathbf{R}coll$S, and that they are not collateral $\mathbf{R}noncoll$S. The question now at once arises how collaterality is related to uncorrelation and we shall take a look at this. We therefore prove the following lemmas and theorem.

Lemma 8.3.3: If $Qsub\mathbf{R}$, $Qsub$S and $\mathbf{R}uncorr$S, then $Quncorr\mathbf{R}$ and $Quncorr$S.

Proof: Let $\mathbf{R}^{Q}$ and S^{Q} be functions and let $\mathbf{R}^{S}$ be sweeping. Suppose that $\mathbf{R}^{Q}$ is not sweeping. Then $Q\mathbf{R}| \mathbf{R}^{S}$ is not sweeping according to 7.3.16. From 7.3.8 follows that $\mathbf{R}^{Q}| Q^{S} \supseteq \mathbf{R}^{S}$. Thus $\mathbf{R}^{S}$ not sweeping according to 7.3.12, and we have proved a contradiction. $\mathbf{R}^{Q}$ is thus sweeping , which implies that $\mathbf{R}uncorr$Q. We have thus proved that if $Qsub\mathbf{R}$, $Qsub$S and $\mathbf{R}uncorr$S, then $Quncorr\mathbf{R}$. Due to symmetry it follows that if $Qsub\mathbf{R}$, $Qsub$S and $\mathbf{R}uncorr$S, then $Quncorr$S. ♦

Lemma 8.3.4: If $Qsub\mathbf{R}$ and $Quncorr\mathbf{R}$, then $\mathbb{C}_{Q}$ is degenerate.

Proof: Suppose that $\mathbb{C}_{Q}$ is non-degenerate. From 7.3.30 then follows that $Qsub\mathbf{R}$

implies **Q**_corr_**R**. But this contradicts **Q**_uncorr_**R**. This shows that $\mathbb{C}_\mathbf{Q}$ is degenerate. ◆

Lemma 8.3.5: If $\mathbb{C}_\mathbf{R}$ is degenerate, then $\mathbb{I}_\mathbf{R}=\mathbb{B}i$.
Proof: Suppose that $\mathbb{I}_\mathbf{R}\neq\mathbb{B}i$. Then there are $A,B\in \mathbb{D}$ and $\varphi\in \mathbb{B}i(A,B)$ such that $\varphi\notin \mathbb{I}_\mathbf{R}(A,B)$. Thus, $\varphi[R(A)]\neq R(B)$. $\varphi[R(A)]\in \mathbb{C}_\mathbf{R}$ and has the same base set as $R(B)$. $\mathbb{C}_\mathbf{R}$ is thus non-degenerate. ◆

Theorem 8.3.6: If **R**_uncorr_**S** then **R**_coll_**S**.
Proof: Suppose that **R**_uncorr_**S** and that **Q**_sub_**R** and **Q**_sub_**S**. Then **Q**_uncorr_**R** according to 8.3.3. From 8.3.4 follows that $\mathbb{D}$ is degenerate and from 8.3.5 that $\mathbb{I}_\mathbf{Q}=\mathbb{B}i$. $\mathbb{B}i$ is thus the only upper bound for $\mathbb{I}_\mathbf{R}$ and $\mathbb{I}_\mathbf{S}$. ◆

The converse of 8.3.6 does not hold.

8.4 Semiranks

As we saw in section 8.2, $\langle\text{Rank } \mathbb{D}, \leq \rangle$ is a complete lattice in the Boolean algebra $\langle \text{Prank } \mathbb{D}, \subseteq \rangle$. The subset Rank $\mathbb{D}$ of Prank $\mathbb{D}$ is characterized by its elements being ranks of relational systems. One might wonder whether it is possible to characterize Rank $\mathbb{D}$ in terms of the Boolean algebra $\langle \text{Prank } \mathbb{D}, \subseteq \rangle$. With this in view, let us consider the set consisting of all F in Prank $\mathbb{D}$ such that the following three conditions are satisfied.

(A1) $F(A,A) \neq \varnothing$

(A2) $F(B,A) = \{\varphi^{-1}\mid \varphi\in F(A,B)\}$

(A3) $\{\psi\circ\varphi\mid \varphi\in F(A,B) \ \& \ \psi\in F(B,C)\} \subseteq F(A,C)$.

It follows from 2.9.1 that if $F\in$ Rank $\mathbb{D}$, then F satisfies (A1) – (A3). However, it does not hold that if $F\in$ Prank $\mathbb{D}$ and F satisfies (A1) – (A3), then $F\in$ Rank $\mathbb{D}$. The conditions are therefore only necessary for F being a rank and not sufficient. To give conditions of the same kind as (A1) – (A3) which are also sufficient seems rather difficult. To see this, we first consider groups of automorphisms.

It is a well-known fact (and easy to prove) that for every relational structure A, the set of automorphisms of A is a group with respect to functional composition o, i.e. $\langle\mathbb{I}(A),o\rangle$ is a group. $\langle\mathbb{I}(A),o\rangle$ is called the group of automorphisms of A or the *automorphism group* of A. It is often convenient to let o be implicit and talk about

$\mathbb{I}(\mathbf{A})$ as the group of automorphisms.

The set of automorphisms of a relational structure is thus a group. What about the converse? In Birkhoff (1946) it is proved that every group is isomorphic to an automorphism group of some algebra. Since an algebra is an relational structure, every group is isomorphic to an automorphism group of some relational structure. It is important in the context of this section to note what does not follow from Birkhoff's result.

$\mathbb{I}(\mathbf{A})$ is of course a subgroup of $\mathbb{B}\mathrm{i}(\mathbf{A})$. Suppose now that $G \subseteq \mathbb{B}\mathrm{i}(\mathbf{A})$ and $\langle G, o \rangle$ is a group. It is now natural to ask whether there always is a structure $\mathbf{A}$ with A as base set such that $\mathbb{I}(\mathbf{A})=G$. From the result in Birkhoff (1946) mentioned above it follows that there is a relational structure $\mathbf{X}$ such that $\langle G, o \rangle$ is isomorphic to $\langle \mathbb{I}(\mathbf{X}), o \rangle$. But it is not certain that $\mathbf{X}=\mathbf{A}$. What conditions a subset G of $\mathbb{B}\mathrm{i}(\mathbf{A})$ must satisfy in order to guarantee the existence of a structure $\mathbf{A}$ with A as base set such that $G=\mathbb{I}(\mathbf{A})$ is as far as I know still an open problem. It is important here that we are only considering relational structures with relations of finite arity. The situation might be quite different if we permit relations of infinite arity.

We now return to relationals. Note that the notion of a rank is in a sense a generalization of the notion of a group of automorphisms. To see this let $\mathbb{D}$ consist of only one set A. The rank of a relational system $\mathbf{R}$ over $\mathbb{D}$ is then the group of automorphisms of $\mathbf{R}(\mathbf{A})$. (The relation between ranks and groups of automorphisms will be discussed from another point of view in chapter 9.) Giving necessary conditions for a pseudorank being a rank would therefore also solve the problem of which conditions a subset G of $\mathbb{B}\mathrm{i}(\mathbf{A})$ should have to satisfy in order for there to exist a structure $\mathbf{A}$ with A as base set such that $G=\mathbb{I}(\mathbf{A})$. Rather than looking for such conditions we will study the set of all pseudoranks over $\mathbb{D}$ which satisfy (A1) – (A3), which we shall call the set of *semiranks* over $\mathbb{D}$. We are primarily interested in how close the notion of a semirank comes to the notion of a rank. If these properties are intimately related then we can perhaps use results about semiranks in our study of ranks. A semirank is from the view of the Boolean algebra of pseudoranks a more "transparent" notion than a rank. and the properties of semiranks may therefore be easier to uncover.

The set of all semiranks over $\mathbb{D}$ is denoted by Srank $\mathbb{D}$. Thus, $F \in \text{Srank} \mathbb{D}$ iff F is a function defined on $\mathbb{D} \times \mathbb{D}$ such that $F(A,B) \subseteq \mathbb{B}\mathrm{i}(A,B)$ and F satisfies $(A1) - (A3)$.

Theorem 8.4.1: Suppose $F \in \text{Srank } D$. Then the following holds for all $A, B, C \in D$:

(i)   If $\varphi, \psi, \omega \in F(A,B)$ then $\omega \circ \psi^{-1} \circ \varphi \in F(A,B)$.

(ii)   If $F(A,B) \neq \varnothing$ and $F(B,C) \neq \varnothing$, then $F(A,C) \subseteq \{\psi \circ \varphi \mid \varphi \in F(A,B)$ & $\psi \in F(B,C)\}$.

(iii) $F(A,A)$ is a group under functional composition $\circ$.

(iv)   If $\varphi \notin F(A,B)$ and $\psi \in F(A,B)$, then $\psi^{-1} \circ \varphi \notin F(A,A)$.

(The straightforward proof is omitted.)

Suppose that $F \in \text{Srank } D$ and that for all $A \in D$, $F(A,A)$ is a group under functional composition $\circ$. It is then possible to construct a relational system which has F as its rank. This is one of the reasons for studying semiranks. One other stems from the the following observation.

If $J \subseteq \text{Srank } D$ then $\cap J \in \text{Srank } D$, i.e. the greatest lower bound of a set of semiranks is a semirank. This follows easily from (A1) – (A3). $\text{Srank } D$ is thus a closure system in $\text{Prank } D$, and since $\langle \text{Prank } D, \subseteq \rangle$ is a complete lattice, this holds also for $\langle \text{Srank}, \subseteq \rangle$. Note that $\text{glb} J$ taken in $\text{Srank } D$ and in $\text{Prank } D$ coincide, but $\text{lub} J$ taken in $\text{Srank } D$ and $\text{Prank } D$ can very well be different. Let $\text{lub}_P J$, $\text{lub}_S J$ and $\text{lub}_R J$ denote the least upper bound of J in $\text{Prank } D$, $\text{Srank } D$ and $\text{Rank } D$. respectively Then it holds that

$$\text{lub}_P J \subseteq \text{lub}_S J \subseteq \text{lub}_R J.$$

Note that the greatest element in $\text{Prank } D$, $\text{Srank } D$ and $\text{Rank } D$ is the same, namely $\mathbb{B}i$.

8.5 On equality preservation and independent realizability of structures

In this section we shall express in another way what has already been said in section 4.8, namely that the notions of transition, subordination and uncorrelation are basically structural. The reason for giving this new account is that it then becomes evident that *sub* is a kind of equality preservation and *uncorr* a kind of independent realizability.

Two set-theoretical structures X and Y are often said to have the same structure or form if they are isomorphic, i.e. if there exists an isomorphism from X to Y. But, as was emphasized in section 4.8, in a certain sense X and Y have the same structure *relative* to a way of identifying elements in X with those of Y. Such an identification can be described as a bijection from X to Y. X and Y have the same structure relative to a bijection φ from X to Y if φ is an isomorphism from **X** to **Y**. It is obvious that **X** and **Y** can have the same structure relative to one bijection from X to Y without having it relative to another.

We study RelS $\mathbb{D}$—i.e. the set of all systems of relationals with $\mathbb{D}$ as their range of definition—for an arbitrary class $\mathbb{D}$ of sets. Construct the class of equivalence classes modulo cardinality generated from $\mathbb{D}$. For each such equivalence class κ, choose a set σ with the same cardinality as the elements in κ and call it the representative of κ. Let r be a mapping which assigns to every set A in $\mathbb{D}$ the representative of the equivalence class to which A belongs. We call r(A) the *reference set* for A and r the *reference mapping* of $\mathbb{D}$.

In the sequel we shall deal a lot with objects of the form $\langle A,\zeta\rangle$, $\langle R(A),\zeta\rangle$ and $\langle X,\zeta\rangle$. For convenience we shall write simply $A\zeta$, $R(A)\zeta$ and $X\zeta$, omitting brackets and commas. I think it is also possible to treat, for example, $A\zeta$ as an object, which is not necessarily an ordered pair but characterized by rules described below regulating its use. However, this is not an important point in the present context and we can be satisfied with the set-theoretical construction of them as ordered pair. Note that the symbol $A\zeta$ is not used in the way which is typical in algebra, where $A\zeta$— when A is a set and ζ a mapping—means $\zeta[A]$.

Define $\mathbb{D}$+ as

$$\{A\zeta \mid A\in\mathbb{D} \ \& \ \zeta\in\mathrm{Bi}(A,r(A))\}.$$

We read $A\zeta$ as "A under ζ" and $\mathbb{D}$+ as *the extended range of definition* of the elements in RelS $\mathbb{D}$.

The characteristic class of a relational is now extended analogously.

$$\mathbb{C}_{R}+ = \{X\xi \mid \exists A\in\mathbb{D}: R(A)\cong X \ \& \ \xi\in\mathrm{Bi}(X,r(A))\}.$$

(Note that $\cong$ means "isomorphic to".) We call $\mathbb{C}_{R}+$ *the extended reference class* of **R**. It is convenient to generalize the reference mapping r so that if X has the same cardinality as a set A belonging to $\mathbb{D}$ then r(X)=r(A) and we call r(X) the reference set for X. $X\xi$ is read "X under ξ".

Let us define a binary relation $\cong$ on $\mathbb{C}_{R}+$ in the following way. For all $X\xi$ and $Y\upsilon$ in $\mathbb{C}_{R}+$

$$X\xi\cong Y\upsilon \ \text{iff} \ \upsilon^{-1}\circ\xi\in\mathbb{I}(X,Y).$$

Note that

$$X\xi\cong Y\upsilon \ \text{iff} \ \xi[X] =\upsilon[Y].$$

We can interpret $X\xi\cong Y\upsilon$ as saying that $X\xi$ has the same structure as $Y\upsilon$ or, alternatively, that X under ξ has the same structure as Y under υ.

For every R in RelSD let $R+$, the *extended* R, be the function defined on $D+$ such that

$$R+(A\zeta) = R(A)\zeta.$$

For every R in RelSD define a binary relation $\cong_R$ on $D+$ in the following way. For all $A\zeta, B\vartheta$ in $D+$

$$A\zeta \cong_R B\vartheta \text{ iff } \vartheta^{-1}o\zeta \in I_R(A,B).$$

Hence,

$$A\zeta \cong_R B\vartheta \text{ iff } \zeta[R(A)] = \vartheta[R(B)].$$

Note that $\cong$ is used with different meanings in different contexts. In particular we distinguish between the use of $\cong$ in $X\cong Y$, $X\xi \cong Y\upsilon$ and $A\zeta \cong_R B\vartheta$. The following relations hold. $A\zeta \cong_R B\vartheta$ iff $R(A)\zeta \cong R(B)\vartheta$, and $R(A)\zeta \cong R(B)\vartheta$ implies that $R(A) \cong R(B)$. Furthermore, $X \cong Y$ implies that there are ξ and υ such that $X\xi \cong Y\upsilon$.

$\cong$ and $\cong_R$ are in a sense equality relations. $X \cong Y$ means that X and Y have the same (or equal) structure. $X\xi \cong Y\upsilon$ means that $X\xi$ has the same (or equal) structure as $Y\upsilon$. $A\zeta \cong_R B\vartheta$ means that $R(A)\zeta$ and $R(B)\vartheta$ have the same (or equal) structure— i.e. with respect to R, $A\zeta$ and $B\vartheta$ have the same structure. We can also say that $A\zeta$ and $B\vartheta$ have equal R-structure.

$\cong$ and $\cong_R$ are equivalence relations. We prove that this holds for $\cong_R$.

Lemma 8.5.1: $\cong_R$ is an equivalence relation.
Proof: (1) $A\zeta \cong_R A\zeta$ since $\zeta^{-1}o\zeta = \iota_A \in I_R(A)$. Hence, $\cong_R$ is reflexive.
(2) $A\zeta \cong_R B\vartheta$ iff $\vartheta^{-1}o\zeta \in I_R(A,B)$ iff $(\vartheta^{-1}o\zeta)^{-1} \in I_R(B,A)$ iff $\zeta^{-1}o\vartheta \in I_R(B,A)$ iff $B\vartheta \cong_R A\zeta$. Hence, $\cong_R$ is symmetric.
(3) Suppose that $A\zeta \cong_R B\vartheta$ and $B\vartheta \cong_R C\eta$. Then $\vartheta^{-1}o\zeta \in I_R(A,B)$ and $\eta^{-1}o\vartheta \in I_R(B,C)$. This implies that $(\eta^{-1}o\vartheta)o(\vartheta^{-1}o\zeta) \in I_R(A,C)$ and thus $(\eta^{-1}o\zeta) \in I_R(A,C)$. From this it follows that $A\zeta \cong_R C\eta$. Hence, $\cong_R$ is transitive. ♦

In the following lemma it is shown that there is another natural way to characterize C_{R+}.

Lemma 8.5.2: $C_{R+} = \{X\xi \mid \exists A\zeta \in D+: \xi \in Bi(X,r(A)) \text{ \& } R(A)\zeta \cong X\xi\}$.
Proof: (I) Suppose $X\xi \in C_{R+}$. Then there is $A \in D$ such that $\xi \in Bi(X,r(A))$ and there is $\varphi \in I(R(A),X)$. $\xi o \varphi \in Bi(A,r(A))$ so $A(\xi o \varphi) \in D+$, and since $\xi^{-1}o(\xi o \varphi) = \varphi$ it

holds that

$$R(A)(\xi o\phi) \cong X\xi.$$

If we let $\zeta = \xi o\phi$ we get $A\zeta \in D+$ and $R(A)\zeta \cong X\xi$.

(II) Suppose there is $A\zeta \in D+$ such that $\xi \in Bi(X,r(A))$ and $R(A)\zeta \cong X\xi$. Then $\xi^{-1}o\zeta \in I(R(A),X)$ which implies that $R(A) \cong X$. Hence, $X\xi \in C_{R+}$. ♦

(8.5.2) shows that it is natural to see C_{R+} as the characteristic class of $R+$.

Some of the motivation for studying $\cong_R$ stems from the following facts, formulated as two theorems.

Theorem 8.5.3: $\cong_R \subseteq \cong_S$ iff $S subR$.

Proof: (I) Suppose that $\cong_R \subseteq \cong_S$ holds. Suppose further that $\phi \in I_R(A,B)$. Let $\zeta \in Bi(A,r(A))$. Hence $\zeta o\phi^{-1} \in Bi(B,r(A))$. Then $A\zeta \cong_R B(\zeta o\phi^{-1})$ since

$$(\zeta o\phi^{-1})^{-1}o\zeta = \phi o\zeta^{-1}o\zeta = \phi .$$

From the assumption it follows that $A\zeta \cong_S B(\zeta o\phi^{-1})$, which implies that

$$(\zeta o\phi^{-1})^{-1}o\zeta \in I_S(A,B)$$

and hence $\phi \in I_S(A,B)$. This proves that $S subR$.

(II) Suppose that $S subR$. Suppose further that $A\zeta \cong_R B\vartheta$. Then $\vartheta^{-1}o\zeta \in I_R(A,B)$. Since $S subR$, $\vartheta^{-1}o\zeta \in I_S(A,B)$ and hence $A\zeta \cong_S B\vartheta$. This proves that $\cong_R \subseteq \cong_S$. ♦

Theorem 8.5.4: $R uncorr S$ iff for every $A\zeta, B\vartheta \in D+$ such that $cardA=cardB$ there is $C\eta \in D+$ such that $A\zeta \cong_R C\eta$ and $B\vartheta \cong_S C\eta$.

Proof: (I) Suppose $R uncorr S$ and let $A\zeta, B\vartheta \in D+$ such that $cardA=cardB$. It follows that $\zeta^{-1}o\vartheta \in Bi(B,A)$. Hence, $(\zeta^{-1}o\vartheta)[S(B)] \in C_S$ and since $R uncorr S$ it follows that $R(A)R^S(\zeta^{-1}o\vartheta)[S(B)]$. Therefore there is, according to 7.2.1, $C \in D$ and $\chi \in Bi(C,A)$ such that $R(A)=\chi[R(C)]$ and $(\zeta^{-1}o\vartheta)[S(B)]=\chi[S(C)]$. Hence, $\zeta[R(A)]=(\zeta o\chi)[R(C)]$ and $\vartheta[S(B)]=(\zeta o\chi)[S(C)]$ and since $\zeta o\chi \in Bi(C,r(A))$ it follows that $A\zeta \cong_R C(\zeta o\chi)$ and $B\vartheta \cong_S C(\zeta o\chi)$. If we choose $\eta=\zeta o\chi$ we get $C\eta \in D+$ and $A\zeta \cong_R C\eta$ and $B\vartheta \cong_S C\eta$.

(II) Suppose that for every $A\zeta, B\vartheta \in D+$ such that $cardA=cardB$ there is $C\eta \in D+$ such that $A\zeta \cong_R C\eta$ and $B\vartheta \cong_S C\eta$. Suppose further that $X \in C_R$ and $Y \in C_S$ and $X=Y$. Then there is $A_0, B_0 \in D$ such that $\phi \in Bi(A_0,X)$ and $\psi \in Bi(B_0,Y)$ such that $\phi[R(A_0)]=X$ and $\psi[S(B_0)]=Y$.

Let $\chi \in Bi(X,r(X))$. Then $(\chi o\phi)[R(A_0)]=\chi[X]$ and $(\chi o\psi)[S(B_0)]=\chi[Y]$. Since

$(\chi o \phi) \in \mathbb{B}i(A_0,r(A_0))$ and $(\chi o \psi) \in \mathbb{B}i(B_0,r(B_0))$ it follows from the assumption that there is $C_0\eta$ such that $A_0(\chi o \phi) \cong_R C_0\eta$ and $B_0(\chi o \psi) \cong_S C_0\eta$. Hence, $(\chi o \phi)R(A_0)=$ $=\eta[R(C_0)]$ and $(\chi o \psi)S(A_0)=\eta[S(C_0)]$. This implies that $X=(\chi^{-1}o\eta)[R(C_0)]$ and $Y=(\chi^{-1}o\eta)[S(C_0)]$, and from this it follows that XR^SY. R^S is thus uncorrelated. ◆

8.5.3 means that *sub* expresses a condition of preservation of equality, more precisely preservation of equality of structure. (8.5.4) shows that *uncorr* is a kind of independent realizability condition. This last point is worth elaborating.

We say that $C\eta$ *realizes* $A\zeta$ for R if $A\zeta \cong_R C\eta$. And by analogy with what holds for $\sim_\alpha$ and $\sim_\beta$ (see section 6.1) we say that $C\eta$ *realizes simultaneously* $A\zeta$ for R and $B\vartheta$ for S if $A\zeta \cong_R C\eta$ and $B\vartheta \cong_S C\eta$. Further, we say that $C\eta$ *realizes* $X\xi$ for R if $X\xi \cong R(C)\eta$. And $C\eta$ *realizes simultaneously* $X\xi$ for R and $Y\upsilon$ for S if $X\xi \cong R(C)\eta$ and $Y\upsilon \cong S(C)\eta$. Finally, by analogy with what holds for $\sim_\alpha$ and $\sim_\beta$ we say that R and S are *independently realizable* iff for all $A\zeta, B\vartheta \in \mathbb{D}^+$ there exists a $C\eta \in \mathbb{D}^+$ such that $C\eta$ realizes simultaneously $A\zeta$ for R and $B\vartheta$ for S. Hence, according to (8.5.4), R and S are *independently realizable* iff Runcorr$S$.

This last proposition motivates the introduction of the *realization class from* R *to* S where R and S are relational systems with the same range of definition $\mathbb{D}$. We denote it by $R{\Downarrow}S$ and define it in the following way:

$$R{\Downarrow}S = \{\langle A\zeta, B\vartheta\rangle \mid \exists C\eta \in \mathbb{D}^+: A\zeta \cong_R C\eta \ \& \ B\vartheta \cong_S C\eta\}$$

Let us say that $A\zeta$ for R and $B\vartheta$ for S are *simultaneously realizable* if there is $C\eta \in \mathbb{D}^+$ such that $C\eta$ realizes simultaneously $A\zeta$ for R and $B\vartheta$ for S. The realization class from R to S is then the set of ordered pairs $\langle A\zeta, B\vartheta\rangle$ such that $A\zeta$ for R and $B\vartheta$ for S are simultaneously realizable. Note that $R{\Downarrow}S$ is the relative product of $\cong_R$ and $\cong_S$, i.e. $(\cong_R|\cong_S)$.

The relation between *uncorr* and the realization class is expressed in the following proposition, which can be regarded as a corollary to 8.5.4.

Corollary 8.5.5: Runcorr$S$ if and only if for all $A\zeta, B\vartheta \in \mathbb{D}^+$,
$$A\zeta(R{\Downarrow}S)B\vartheta \text{ iff card}A=\text{card}B.$$

The next proposition illustrates the relation between the notion of *sub* and of realization class.

Proposition 8.5.6: S*sub*R if and only if for all $A\zeta, B\vartheta \in \mathbb{D}^+$,

$$A\zeta(R\Downarrow S)B\vartheta \quad \text{implies} \quad A\zeta\cong_S B\vartheta.$$

Proof: (I) Suppose $S\,sub\,R$. Suppose further that $A\zeta(R\Downarrow S)B\vartheta$. Then there is $C\eta\in\mathbb{D}^+$ such that $A\zeta\cong_R C\eta$ and $B\vartheta\cong_S C\eta$. Since $S\,sub\,R$ it follows from 8.5.3 that $\cong_R\subseteq\,\cong_S$ and hence $A\zeta\cong_S C\eta$. From this and $B\vartheta\cong_S C\eta$ it follows that $A\zeta\cong_S B\vartheta$, since $\cong_S$ is an equivalence relation .

(II) Suppose the second part of the equivalence in the proposition. Suppose further that $A\zeta\cong_R B\vartheta$. Since $\cong_S$ is an equivalence relation it holds that $B\vartheta\cong_S B\vartheta$. Hence $A\zeta(R\Downarrow S)B\vartheta$ and from the assumption follows $A\zeta\cong_S B\vartheta$. This shows that $\cong_R\subseteq\,\cong_S$. According to 8.5.3 this implies that $S\,sub\,R$. ♦

Note that if $F\in Prank\mathbb{D}$, then $\cong_F$ defined by
$$A\zeta\cong_F B\vartheta \quad \text{iff} \quad \vartheta^{-1}o\zeta\in F(A,B)$$
is a relation in $\mathbb{D}^+$. It is easy to see $\cong_F$ is an equivalence relation iff $F\in Srank\mathbb{D}$. This sheds some light on the notion of a semirank.

CHAPTER 9

ISOMORPHIC MAPPINGS AND INVARIANCE

9.0 Introduction

A relational is a kind of generalization of a relation in the ordinary sense of a set of ordered n-tuples. One might therefore ask if the notion of an isomorphism for a relation in the ordinary sense can be generalized to relationals. In section 9.2 we will see how this can be done. Given this notion of an isomorphism for a relational it is a straightforward matter to say what is meant by an automorphism for a relational. As a preliminary to the study of isomorphisms for relationals, the notion of a mapping is introduced in section 9.1. In section 9.3 we study invariance under the set of automorphisms for a relational and see how this notion is related to *subordination*. The idea of stability can be applied in a natural way to automorphisms for relationals, and invariance under stable automorphisms will be studied in section 9.4 It turns out that invariance under stable automorphisms gives rise to a notion of dependence of different character to that of subordination because it is not domainwise. We shall be primarily interested in a notion which is somewhat stronger than invariance under stable automorphisms, to be called *global subordination*, by contrast with the usual *local* subordination. (The notion of global subordination and the theorems 9.4.4 and 9.4.5 were suggested to me by Sten Lindström.)

9.1 Mappings

We will generalize the notion of an isomorphism for a relation in the ordinary sense of a set of ordered n-tuples in the next section so that it can be applied to relationals. To avoid confusion with the traditional notion of isomorphism we shall call the generalized notion *isomorphic mapping*. This implies a special use of the term "mapping" and in this section we define what is to be meant by it here.

Suppose that M and M' are non-empty classes whose elements are non-empty sets. A *mapping* on M into M' is an ordered pair $<j,J>$ where j is a function on M into M' and J a function with M as domain and such that for all $A \in M$, J(A) is a function on A into j(A). Thus,

$$j: M \to M'$$

and for all $A \in M$

$$J(A): A \to j(A).$$

We call M the the *source* of the mapping $<j,J>$ and call $\{J(A)[A] \mid A \in M\}$ the *target* of $<j,J>$. A mapping $<j,J>$ on M into M' is *surjective* if j is a function on M onto M' and for all $A \in M$, J(A) is a function on A onto j(A). If $<j,J>$ is a surjective mapping on M into M', then the target of $<j,J>$ equals M'. A mapping $<j,J>$ on M into M' is *second-half-surjective* if for all $A \in M$, J(A) is a function on A onto j(A). A mapping $<j,J>$ on M into M' is *injective* if the inverse of j is a function and for all $A \in M$ the inverse of J(A) is a function. A mapping $<j,J>$ on M into M' is *bijective* if it is surjective and injective. It is easy to see that a mapping $<j,J>$ on M into M' is bijective iff $j \in \text{Bi}(M,M')$ and for all $A \in M$, $J(A) \in \text{Bi}(A,j(A))$. We denote the set of bijective mappings on M into M' by $\text{Bim}(M,M')$. The set of bijective mappings on M into M is denoted $\text{Bim}(M)$ rather than $\text{Bim}(M,M)$, and if $<j,J> \in \text{Bim}(M)$ we say that that $<j,J>$ is a bijective mapping on M.

The *identity mapping* of the class M of non-empty sets is $<\iota,\Upsilon>$ where ι is the identity function on M and for all $A \in M$, $\Upsilon(A)$ is the identity function on A, i.e. for all $A \in M$ and all $a \in A$,

$$\iota(A)=A \quad \text{and} \quad \Upsilon(A)(a)=a .$$

We now define the composition operation for mappings. Let $<j_1,J_1>$ be a mapping on M_1 and $<j_2,J_2>$ a mapping on M_2. The *composition* of $<j_1,J_1>$ and $<j_2,J_2>$, which we denote by

$$<j_2,J_2> \text{\textcopyright} <j_1,J_1>,$$

is the ordered pair $<j,J>$ where

$$j = j_2 \circ j_1$$

and for all $A \in M$,

$$J(A) = J_2(j_1(A)) \circ J_1(A),$$

where $M = \{A \in M_1 \mid j_1(A) \in M_2\}$.

Note that $j: M \to M_2$ and for all $A \in M$,

$$J(A) : A \to (j_2 o j_1)(A).$$

The last assertion holds since

$$J_1(A) : A \to j_1(A)$$

and

$$J_2(j_1(A)) : j_1(A) \to j_2(j_1(A)).$$

Thus, $<j,J>$ is a mapping with $\{A \in M_1 \mid j_1(A) \in M_2\}$ as the source and with

$$\{ J_2(j_1(A))[J_1(A)[A]] \mid j_1(A) \in M_2\}$$

as the target.

Note that $\{A \in M_1 \mid j_1(A) \in M_2\}$ can very well be empty. In such a case j and J are both empty and we call $<j,J>$ the empty mapping. If $\{A \in M_1 \mid j_1(A) \in M_2\}$ is non-empty, then j, J and the target are non-empty and we say that $<j,J>$ is non-empty.

Note that if $<j,J>$ is a mapping on M into M' and $<\iota, \Upsilon>$ is the identity mapping on M', then

$$<\iota, \Upsilon> \odot <j,J> = <j,J>.$$

And if $<\iota, \Upsilon>$ is the identity mapping of M, then

$$<j,J> \odot <\iota, \Upsilon> = <j,J>.$$

Composition of functions is an associative operation. We expect of course composition of mappings to be associative, too. That this is the case is proved as Lemma 9.1.1.

Theorem 9.1.1: $\odot$ is associative, i.e.

$$<j_3,J_3> \odot (<j_2,J_2> \odot <j_1,J_1>) = (<j_3,J_3> \odot <j_2,J_2>) \odot <j_1,J_1>.$$

Proof: $<j_2,J_2> \odot <j_1,J_1> = <j_4,J_4>$ where $j_4 = j_2 o j_1$ and $J_4(A) = J_2(j_1(A)) o J_1(A)$ for all A in $M_4 = \{A \in M_1 \mid j_1(A) \in M_2\}$. $<j_3,J_3> \odot <j_4,J_4> = <j,J>$ where $j = j_3 o j_4$ and $J(A) = J_3(j_4(A)) o J_4(A)$ for all A in

$$M = \{A \in M_4 \mid j_4(A) \in M_3\} = \{A \in M_1 \mid j_1(A) \in M_2 \ \& \ j_2(j_1(A)) \in M_3\}.$$

Thus, $j = j_3 o j_2 o j_1$ and $J(A) = J_3(j_2(j_1(A))) o J_2(j_1(A)) o J_1(A)$ for all $A \in M$.

$<j_3,J_3> \odot <j_2,J_2> = <j_5,J_5>$ where $j_5 = j_3 o j_2$ and $J_5(A) = J_3(j_2(A)) o J_2(A)$ for all A in $M_5 = \{A \in M_2 \mid j_2(A) \in M_3\}$. $<j_5,J_5> \odot <j_1,J_1> = <j',J'>$ where $j' = j_5 o j_1$ and $J'(A) = = J_5(j_1(A)) o J_1(A)$ for all A in

$$M' = \{A \in M_1 \mid j_1(A) \in M_5\} = \{A \in M_1 \mid j_1(A) \in M_2 \ \& \ j_2(j_1(A)) \in M_3\}$$

i.e. $M' = M$. Hence, $j' = j_3 o j_2 o j_1$ and $J'(A) = J_3(j_2(j_1(A))) o J_2(j_1(A)) o J_1(A)$ for all $A \in M$. We therefore have $j = j'$ and $J(A) = J'(A)$ for all $A \in M$. $\blacklozenge$

We now define the inversion operations for mappings. If $<j,J>$ is a second-half-surjective mapping on M into M', then the *inverse mapping,* denoted by $<j,J>^i$, is the ordered pair $<j^{-1},J^*>$ where j^{-1} is the inverse of j and for all A in M'

$$J^*(A) = J(j^{-1}(A))^{-1}.$$

The source of $<j,J>^i$ is the target of $<j,J>$. Some simple properties of inverse mappings are stated in the following lemma.

Lemma 9.1.2: Suppose $<j,J>$ is a bijective mapping on M into M' and $<j',J'>$ a bijective mapping on M' into M". Then

(i) $<j,J>^i$ is a bijective mapping on M' into M.

(ii) $(<j,J>^i)^i = <j,J>$.

(iii) $<j,J>^i o<j,J>$ is the identity mapping on M and $<j,J>o<j,J>^i$ is the identity mapping on M'.

(iv) $<j',J'> © <j,J>$ is a bijective mapping on M into M".

Proof: (i) Suppose$<j,J>\in \mathbb{B}im(M,M')$. Then $j\in \mathbb{B}i(M,M')$ and thus $j^{-1}\in \mathbb{B}i(M',M)$. Let $<j,J>^i=<j^{-1},J^*>$ and $A\in M'$. Then $J^*(A)=J(j^{-1}(A))^{-1}$. $J(j^{-1}(A))\in \mathbb{B}i(j^{-1}(A),A)$ and thus $J(j^{-1}(A))^{-1}\in \mathbb{B}i(A,j^{-1}(A))$. This shows that $<j,J>^i\in \mathbb{B}i(M',M)$.

(ii) $(<j,J>^i)^i = <(j^{-1})^{-1},J">$ where for all A in M $J"(A) = J^*(j^{-1}(A))^{-1}$ and $J^*(A) = J(j^{-1}(A))^{-1}$ for all A in M'. Since j is a bijection and $J(X)$ a bijection for all $X\in M$ it follows that for all $A\in M$,

$$J"(A) = (J(j^{-1}(j^{-1}(A)))^{-1})^{-1} = J(A).$$

Thus, $<(j^{-1})^{-1},J"> = <j,J>$.

(iii)$<j,J>^i o<j,J>=<j_0,J_0>$ where $j_0=j^{-1}oj=\iota_M$ and, since j and $J(A)$ are bijections,

$$J_0(A) = J^*(j(A)) o J(A) = J(j^{-1}(j(A)))^{-1} o J(A) = J(A)^{-1} o J(A) = \iota_A.$$

Analogously, it is proved that $<j,J>o<j,J>^i$ is the identity mapping on M'.

(iv) $<j',J'> © <j,J> = <j^°,J^°>$ such that $j^° = j'oj$ and for all $A\in M$,

$$J^°(A) = J'(j(A))oJ(A).$$

Thus, $j^°\in \mathbb{B}i(M,M")$. $J(A)\in \mathbb{B}i(A,j(A))$ and $J'(j(A))\in \mathbb{B}i(j(A),j'(j(A)))$, which implies that $J^°(A)\in \mathbb{B}i(A,j'(j(A)))$. This shows that $<j^°,J^°>\in \mathbb{B}i(M,M")$. ♦

We now introduce two special kinds of mappings which we will make use of in the sequel. A mapping $<j,J>$ on M into M' such that j is the identity mapping on M is called a *conservative mapping.* A conservative mapping is thus always of the form $<\iota_M,J>$, and therefore for all $A\in M$, $J(A):A\to A$.

A mapping <j,J> on M into M' such that for all A,B∈ M,

$$J(A){\restriction}B = J(B){\restriction}A$$

is called a *stable mapping*. A mapping <j,J> is thus stable iff for all A,B∈ M and for all x in A∩B,

$$J(A)(x) = J(B)(x)$$

Suppose <j,J> is a mapping of M into M'. Define

$$J\# = \{\langle a,b\rangle \in (\cup M \times \cup M') \mid \exists A \in M : \exists a \in A \ \& \ J(A)(a)=b\}$$

where ∪M={a | ∃A∈ M: a∈ A}. Note that J# has ∪M as domain.

Lemma 9.1.3: Suppose <j,J> is a mapping on M into M'. Then <j,J> is stable iff J# is a function.

Proof: (I) Suppose <j,J> is stable. Suppose further that ⟨a,b⟩,⟨a,b'⟩∈ J#. Then there are A,A'∈ M such that a∈ A and J(A)(a)=b while a∈ A' and J(A')(a)=b'. Since J is stable, J(A)(a)=J(A')(a), which implies b=b'. Thus, J# is a function.

(II). Suppose that J is not stable. Then there are A,B∈ M and a∈ A∩B such that J(A)(a)≠J(B)(a). Since ⟨a,J(A)(a)⟩∈ J# and ⟨a,J(B)(a)⟩∈ J# then J# is not a function. Thus, if J# is a function then <j,J> is stable. ♦

Lemma 9.1.4: Suppose <j,J> is a mapping on M into M'. If <j,J> is stable then for all A∈ M, J(A)=J#↾A.

Proof: Suppose <j,J> is stable. Then, according to 9.1.3, J# is a function. Let a∈ A∈ M. Then ⟨a, J(A)(a)⟩∈ J#, i.e. J#(a)=J(A)(a). Since this holds for all a∈ A, J#↾A=J(A). ♦

The inverse of a stable mapping need not be stable. The following example shows this. Let M={A,B}, A={a,c} and B={b}; M'{A',B'}, A'={a',b'} and B={b'}; j={⟨A,A'⟩,⟨B,B'⟩} and

$$J = \{\langle A,\{\langle a,a'\rangle,\langle c,b'\rangle\}\rangle, \ \langle B,\langle b,b'\rangle\}\rangle\}.$$

<j,J>∈ Bim(M,M') and <j,J> is obviously stable. Since <j,J> is bijective the inverse of <j,J> exists. <j,J>^{i}=<j^{-1},J*> where

$$J^*(A') = J(j^{-1}(A'))^{-1} = J(A)^{-1}$$
$$J^*(B') = J(j^{-1}(B'))^{-1} = J(B)^{-1}.$$

is not stable since b'∈ A' and b'∈ B' and J*(A')(b')=c and J*(B')(b')=b. Note that J#={⟨a,a'⟩,⟨c,b'⟩,⟨b,b'⟩} so the inverse of J# does not exist even though <j,J> is a bijective mapping.

The above example motivates, as will soon be obvious, the following definition. Suppose <j,J> is a mapping on M into M'. We say that <j,J> is *global* if it is stable and for all A,B∈ M,

$$J(A)[A∩B] = J(A)[A] ∩ J(B)[B].$$

(The notion of global was suggested to me by Sten Lindström.) In the example above, J(A)[A∩B]=J(A)[∅] but J(A)[A] ∩ J(B)[B] = {b'} , so <j,J> is not global.

Note that if <j,J> is stable then for all A,B∈ M,

$$J(A)[A∩B] ⊆ J(A)[A] ∩ J(B)[B].$$

To see this, suppose that x∈ J(A)[A∩B]. Then there is a∈ A∩B such that J(A)(a)=x. Since <j,J> is stable, J(B)(a)=J(A)(a)=x. Thus x∈ J(A)[A] ∩ J(B)[B]. This shows that J(A)[A∩B] ⊆ J(A)[A] ∩ J(B)[B]. It therefore holds that <j,J> is global iff it is stable and for all A,B∈ M,

$$J(A)[A∩B] ⊇ J(A)[A] ∩ J(B)[B].$$

Lemma 9.1.5: Suppose <j,J> is a mapping on M into M'. If J# is a one-to-one function then <j,J> is global.

Proof: Since J# is a function, it follows that <j,J> is stable according to 9.1.3. Suppose so that x∈ J(A)[A] ∩ J(B)[B]. Then there is a∈ A such that J(A)(a)=x and b∈ B such that J(B)(b)=x. Thus ⟨a,x⟩,⟨b,x⟩∈ J#. Since J# is a one-to-one function this implies that a=b, and therefore a∈ A∩B and x∈ J(A)[A∩B]. We have proved that J(A)[A] ∩ J(B)[B] ⊆ J(A)[A∩B]. ♦

Lemma 9.1.6: Suppose <j,J> is a mapping on M into M' such that for all A∈ M, J(A) is one-to-one. If <j,J> is global then J# is a one-to-one function.

Proof: Suppose that <j,J> is global. Then <j,J> is stable, so J# is a function according to 9.1.3.Suppose ⟨a,x⟩,⟨b,x⟩∈ J#. Then there are A,B∈ M such that a∈ A,b∈ B and J(A)(a)=x, J(B)(b)=x. Thus x∈ (J(A)[A] ∩ J(B)[B]), and since <j,J> is global, x∈ J(A)[A∩B]. Then there is c∈ A∩B such that J(A)(c)=x. Since <j,J> is stable, J(B)(c)=J(A)(c)=x.. According to the assumption, J(A) and J(B) are one-to-one. Hence, a=c=b. This shows that J# is one-to-one. ♦

Lemma 9.1.7: Suppose <j,J> is a bijective mapping on M into M'. If <j,J> is global then J# is a one-to-one correspondence on ∪M onto ∪M'.

Proof: According to 9.1.6 it remains to prove that J# is onto ∪M'. Let a'∈ ∪M'. Then there is A'∈ M' such that a'∈ A'. Since <j,J> bijective there is A∈ M such that j(A)=A' and a∈ A such that J(A)(a)=a'. Thus ⟨a,a'⟩∈ J#. This shows that J# is a function onto ∪M'. ◆

Lemma 9.1.8: Suppose <j,J> is a mapping on M into M' and J# is a one-to-one correspondence on ∪M onto ∪M'. If j onto M' and for all A∈ M, j(A)=J(A)[A], then <j,J>∈ Bim(M,M').
Proof: (I) We show first that j∈ Bi(M,M'). Suppose that j(A)=j(B), i.e. J(A)[A]=J(B)[B]. Since J# is one-to-one, it follows from 9.1.5 and 9.1.4 that J(A)=J#⌈A and J(B)=J#⌈B and thus A=B. According to the assumption, j onto M'. (II) We then show that for all A∈ M, J(A)∈ Bi(A,j(A)). According to the assumption, j(A)=J(A)[A]. Thus J(A) is onto j(A). Since J# is one-to-one and J(A)=J#⌈A, J(A) is one-to-one. ◆

Lemma 9.1.9: Suppose <j,J> is a bijective mapping on M into M'. Then (J*)#= =(J#)⁻¹ where J* is defined by <j,J>ⁱ=<j⁻¹,J*>.
Proof: (I) Suppose ⟨a,b⟩∈ J#. Then there is A∈ M such that a∈ A and J(A)(a)=b. Since <j,J> is bijective, J*(j(A))=J(j⁻¹(j(A)))⁻¹=J(A)⁻¹. Thus J*(j(A))(b)=a which implies that ⟨b,a⟩∈ (J*)#.
(II) Suppose now that ⟨a,b⟩∈ (J*)#. Then there is A∈ M' such that a∈ A and J*(A)(a)=b. J*(A)=J(j⁻¹(A))⁻¹. Thus J(j⁻¹(A))(b)=a. Since j⁻¹(A)∈ M, ⟨b,a⟩∈ J#. ◆

Lemma 9.1.10: Suppose <j,J> is a bijective mapping on M into M'. Then <j,J> is global iff <j,J>ⁱ is global.
Proof: (I) If <j,J> is bijective and global then, according to 9.1.7, J# is a one-to-one correspondence on ∪M onto ∪M'. Thus (J#)⁻¹ is a one-to-one correspondence on ∪M' onto ∪M. By 9.1.5, this also holds for (J*)#, where J* is defined by <j,J>ⁱ=<j⁻¹,J*>. Hence, according to 9.1.2(i) and 9.1.8, <j⁻¹,J*> is global.
(II) Suppose <j,J>ⁱ is global. <j,J>ⁱ∈ Bim(M',M) according to 9.1.2(i), since <j,J>∈ Bim(M,M'). Thus (J*)# is a one-to-one correspondence on ∪M' onto ∪M. From this it follows, according to 9.1.9, that J# is a one-to-one correspondence on ∪M onto ∪M'. Hence, <j,J> is global according to 9.1.8. ◆

Let us denote the set of all global bijective mappings on M into M' by GBim(M,M').

Theorem 9.1.11: Let $K \subseteq \mathbb{B}i(\cup M, \cup M')$ such that if $F \in K$ then $F[A] \in M'$ for all $A \in M$ and $F^{-1}[A'] \in M$ for all $A' \in M'$. And let # be the function defined for mappings by the rule

$$\#(<j,J>) = J\#.$$

Then # is a one-to-one function on $\mathbb{GB}im(M,M')$ onto K.

Proof: (I) We first show that # is a function on $\mathbb{GB}im(M,M')$ into K. If $<j,J> \in \mathbb{GB}im(M,M')$ then $J\# \in \mathbb{B}i(\cup M, \cup M')$ according to 9.1.7. And according to 9.1.4, $J\#[A]=j(A)$. Therefore $J\#[A] \in M'$ if $A \in M$. Suppose $A' \in M'$. According to 9.1.9, $(J\#)^{-1}[A']=(J*)\#[A']$ where $J*$ is defined by $<j,J>^i=<j^{-1},J*>$. $<j^{-1},J*> \in \mathbb{GB}im(M',M)$ according to 9.1.2(i) and 9.1.10. This implies that $J*(A')[A']= =(J*)\#[A']$ according to 9.1.4, and thus $j^{-1}(A')=(J*)\#[A']$. Since $j^{-1}(A') \in M$, it follows that $(J\#)^{-1}[A'] \in M$.

(II) We now show that # is onto K. Let $F \in K$. Define for all $A \in M$, $j(A)=F[A]$ and $J(A)=F^\frown A$. Then $j(A) \in M'$ since $F[A] \in M'$ according to the assumption. And $J(A)[A]=j(A)$ since $(F^\frown A)[A]=F[A]=j(A)$. $<j,J>$ is thus a mapping on M into M'. Suppose $A' \in M'$. Since $F \in K$, $F^{-1}[A'] \in M$. Since F is one-to-one, $F[F^{-1}[A']]=A'$, and thus $j(F^{-1}[A'])=A'$. This shows that j is onto M'. From 9.1.8 and 9.1.5 it follows that $<j,J> \in \mathbb{GB}im(M,M')$. We now prove that $J\#=F$. Suppose that $\langle a,b \rangle \in J\#$. Then there is $A \in M$ such that $J(A)(a)=b$. Thus $F(a)=b$. Suppose now that $F(a)=b$. Then there is $A \in M$ such that $a \in A$ and $(F^\frown A)(a)=b$. Thus $J(A)(a)=b$ and $\langle a,b \rangle \in J\#$. We have thus proved that # on $\mathbb{GB}im(M',M)$ is onto K.

(III) We prove now that # is one-to-one. Suppose $<j,J>,<j',J'> \in \mathbb{GB}im(M,M')$ and $\#(<j,J>)=\#(<j',J'>)$ Suppose further that $<j,J> \neq <j',J'>$. Then $j \neq j'$ or $J \neq J'$. Since $<j,J>,<j',J'> \in \mathbb{B}im(M,M')$ it follows that $j \neq j'$ implies $J \neq J'$. Thus there are $A \in M$ and $a \in A$ such that $J(A)(a) \neq J'(A)(a)$. Let $J(A)(a)=x$. Then $\langle a,x \rangle \in J\#$. Since $J\#=J'\#$, there is $B \in M$ such that $a \in B$ and $J'(B)(a)=x$. Since J' is global it is also stable, and therefore $J'(A)(a)=J'(B)(a)$. This implies a contradiction, which shows that $<j,J>=<j',J'>$. ♦

The theorem shows that a global bijective mapping $<j,J>$ can be represented by $J\#$ in a unique way. The proof also shows how, given an element F in K, a global bijective mapping can be constructed, namely by letting $j(A)=F[A]$ and $J(A)=F^\frown A$.

In the next lemma, we state some results for composition of stable and global mappings.

Lemma 9.1.12: Suppose $<j,J>$ is a mapping on M into M' and $<j',J'>$ a mapping on M' into M''. If $<j,J>$ and $<j',J'>$ are stable, then $<j',J'>©<j,J>$ is stable. If $<j,J>$ and $<j',J'>$ are second-half-surjective and global, then $<j',J'>©<j,J>$ is global.

Proof: Let $<j',J'>©<j,J>=<j°,J°>$. (I) Suppose that $<j,J>$ and $<j',J'>$ are stable. Suppose further that $x \in A \cap B$. Since $<j,J>$ is stable, $J(A)(x)=J(B)(x)$. And since $<j,J>$ is stable, $J'(j(A))(J(A)(x))=J'(j(B))(J(B)(x))$. Thus

$$J°(A)(x) = J'(j(A))(J(A)(x)) = J'(j(B))(J(B)(x)) = J°(B)(x).$$

This shows that $<j°,J°>$ is stable.

(II) Suppose that $<j,J>$ and $<j',J'>$ are second-half-surjective and global. As was just shown, $<j°,J°>$ is stable. Suppose now that $A,B \in M$. Then

$$J°(A)[A \cap B] = (J'(j(A)) o J(A))[A \cap B] =$$
$$= J'(j(A))[J(A)[A \cap B]] = J'(j(A))[J(A)[A] \cap J(B)[B]] =$$
$$= J'(j(A))[j(A) \cap j(B)] = J'(j(A))[j(A)] \cap J'(j(B))[j(B)] =$$
$$= J'(j(A))[J(A)[A]] \cap J'(j(B))[J(B)[B]] = J°(A)[A] \cap J°(B)[B].$$

This shows that $J°$ is global. ♦

A mapping $<j,J>$ on M into M' is an ordered pair of functions; j is a function on M into M' and for all $A \in M$, $J(A)$ is a function on A into $j(A)$. We are used to thinking that mappings have arguments and it is therefore natural to ask what are the arguments for $<j,J>$? The answer is simple. If $A \in M$ and $a \in A$ then (A,a) is an argument for $<j,J>$. The set of arguments for $<j,J>$ is thus $\{(A,a) \mid a \in A \in M\}$. The application of $<j,J>$ to the argument (A,a) gives $(j(A),J(A)(a))$ as the result. The set of values of $<j,J>$ is $\{(j(A),J(A)(a)) \mid a \in A \in M\}$. We can therefore regard the mapping $<j,J>$ on M into M' as a function on $\{(A,a) \mid a \in A \in M\}$ into $\{(A,a) \mid a \in A \in M'\}$. Then the following question naturally arises: What is the relation between composition and inversion of mappings on one hand and composition and inversion of functions on the other? We shall now have a look at this.

Suppose $<j_1,J_1>$ is a mapping on M_1 and $<j_2,J_2>$ a mapping on M_2 and $a \in A \in M_1$ and $j(A) \in M_2$. If $a \in A \in M_1$ and $j_1(A) \in M_2$ then

$$(<j_2,J_2> o <j_1,J_1>)(A,a)) = <j_2,J_2>(<j_1,J_1>(A,a)) = <j_2,J_2>(j_1(A),J_1(A)(a)) = (j_2(j_1(A)),$$
$$J_2(j_1(A))(J_1(A)(a)) = ((j_2 o j_1)(A), (J_2(j_1(A) o J_1(A))(a)) = (<j_2,J_2> © <j_1,J_1>)(A,a).$$

This shows that if mappings are regarded as functions, then composition of mappings coincides with composition of functions.

Suppose $<j,J>$ is a bijective mapping with M_1 as source and suppose $a \in A \in M_1$.

Since $\langle j,J\rangle(A,a)=(j(A),J(A)(a))$ we have $\langle j,J\rangle^i(j(A),J(A)(a))=(A,a)$, which implies that $\langle j,J\rangle^i(B,b)=(j^{-1}(B), J(j^{-1}(B))^{-1}(b))$ which is also exactly what we expect from the definition of inversion of bijective mappings.

At this point it is important to note the following. If $\langle j,J\rangle$ is an injective mapping then the inverse of $\langle j,J\rangle$ regarded as a function exists, but the converse does not hold. This implies that $\langle j,J\rangle$ as a function can be a one-to-one correspondence while $\langle j,J\rangle$ might not be a bijective mapping. The following simple examples shows this.

Let $M=\{\{x\},\{y\}\}$, $M'=\{\{x',y'\}\}$, $j=\{\langle\{x\},\{x',y'\}\rangle,\langle\{y\},\{x',y'\}\rangle\}$, $J(\{x\})=\{\langle x,x'\rangle\}$ and $J(\{y\})=\{\langle y,y'\rangle\}$. Hence, the set of arguments of $\langle j,J\rangle$ is $\{(\{x\},x),(\{y\},y)\}$ and the set of values of $\langle j,J\rangle$ is $\{(\{x',y'\},x'),(\{x',y'\},y')\}$. And $\langle j,J\rangle(\{x\},x)=(\{x',y'\},x')$ and $\langle j,J\rangle(\{y\},y)=(\{x',y'\},y')$. $\langle j,J\rangle$ is a one-to-one function while $\langle j,J\rangle$ is not an injective mapping since the inverse of j does not exist. $\langle j,J\rangle$ is a one-to-one correspondence on $\{(A,a)\mid a\in A\in M\}$ onto $\{(A,a)\mid a\in A\in M'\}$ while $\langle j,J\rangle$ does not belong to $\mathbb{Bim}(M,M')$.

9.2 Isomorphic mappings

Two mappings $\langle j,J\rangle$ and $\langle j',J'\rangle$ are equal, which is denoted $\langle j,J\rangle=\langle j',J'\rangle$, iff $j=j'$ and $J=J'$. In general, j is not redundant in a mapping $\langle j,J\rangle$. It holds generally that $J(A)[A]\subseteq j(A)$ but we can have $J(A)[A]=J(B)[B]$ although $j(A)\neq j(B)$. Therefore $\langle j,J\rangle$ and $\langle j',J'\rangle$ can be different even if $J=J'$. However, if $\langle j,J\rangle$ is a second-half-surjective mapping on M into M', i.e. for all $A\in M$ $J(A)$ is onto $j(A)$, then $j(A)=J(A)[A]$ for all $A\in M$. If $\langle j,J\rangle$ and $\langle j',J'\rangle$ are second-half-surjective and $J=J'$, then $j=j'$ so $\langle j,J\rangle=\langle j',J'\rangle$. This means that we can for a second-half-surjective mapping $\langle j,J\rangle$ omit j and denote it simply by J. If we do this, we can recover j according to the rule $j(A)=J(A)[A]$ for all $A\in M$. When a mapping is denoted by J, it is important to distinguish between J as a function and J as a mapping; let us take a look at this.

Suppose $\langle j,J\rangle$ is a mapping on M into M'. Then J is a function with M as domain and such that $J(A)$ is a function on A into $j(A)$. $\langle j,J\rangle$ has M as source and $\{J(A)[A]\mid A\in M\}$ as target. The inverse of J takes functions as arguments and sets as values. The inverse of $\langle j,J\rangle$ is a mapping with the target of $\langle j,J\rangle$ as source and it is denoted by $\langle j,J\rangle^i$. If $\langle j',J'\rangle$ is a mapping on M' into M'', then the composition of $\langle j,J\rangle$ and $\langle j',J'\rangle$ is a mapping with $\{A\in M\mid j(A)\in M'\}$ as source and it is denoted by $\langle j',J'\rangle\copyright\langle j,J\rangle$. The composition of the functions J and J' is not defined since the range of J consists of functions and the domain of J' consists of sets. If <j,J> is a second-half-surjective mapping and we want to denote it by just J, we have to remember the difference between the range and target of J, between the functional inverse and the mapping inverse and between functional composition and mapping composition. J^i is something quite different from J^{-1} and J'©J something quite different from J'oJ. We have that

$$(J'©J)(A) = J'(j(A))oJ(A)$$

and

$$J^i(A)=J(j^{-1}(A))^{-1}.$$

In this section we are mostly interested in bijective mappings which are second-half-surjective and we will therefore use the simplified notation outlined above. A bijective mapping will thus be denoted by a capital letter, and we follow the following conventions. If a mapping is denoted by J, then the corresponding small letter, i.e. j, is the function defined by the rule j(A)=J(A)[A]. We call j *the first half* of J. In cases when a bijective mapping has a complicated name, it is sometimes convenient to denote the first half of the bijection by a special letter. The first half of (J'©J) might for example be denoted by f where f(A)=(J'©J)(A)[A]. The identity mapping <ι,Υ> of the class M will be denoted by Υ. If A∈ M then Υ(A) is the identity function on A. Note that J∈ Bim(M,M') iff M'={j(A) | A∈ M}, j(A)≠j(B) if A≠B and for all A∈ M, J(A)∈ Bi(A,j(A)).

We now prove the anticipated result that Bim(M,M') constitutes a group.

Theorem 9.2.1: ⟨ Bim(M),©,^{i}.Υ⟩ is a group.
Proof: © is an operation in Bim(M) since if J,J'∈ Bim(M) then, according to 9.1.2(iv), (J'©J)∈ Bim(M). According to 9.1.1, © is an associative operation. If J∈ Bim(M) then, according to 9.1.2(i), J^i∈ Bim(M). Υ is the identity (or neutral) element. ♦

We now introduce the important notion of an isomorphic mapping. An *isomorphic mapping* of the system of relationals **R** to the system of relationals **S** is a bijective mapping J on $\mathbb{D}_R$ into $\mathbb{D}_S$ such that for all A∈$\mathbb{D}_R$,

$$J(A)∈\mathbb{I}(R(A),S(j(A))).$$

We let Im(**R,S**) denote the set of isomorphic mappings of **R** to **S**. An isomorphic mapping of **R** to **R** is called an *automorphic mapping* of **R**. This means that J is an automorphic mapping of R iff j∈ Bi($\mathbb{D}_R$) and for all A∈$\mathbb{D}_R$, J(A)∈$\mathbb{I}_R$(A,j(A)).

The set of automorphic mappings of R is denoted by $\text{Im}(R)$ rather than $\text{Im}(R,R)$.

Lemma 9.2.2: If $J \in \text{Im}(Q,R)$ and $J' \in \text{Im}(R,S)$, then $(J' \circledcirc J) \in \text{Im}(Q,S)$.
Proof: It is obvious that $(J' \circledcirc J) \in \text{Bim}(\mathbb{D}_Q, \mathbb{D}_S)$. For all $A \in \mathbb{D}_Q$, $J(A) \in$
$\in \text{I}(Q(A),R(j(A)))$ and $J'(j(A)) \in \text{I}(R(j(A)),S(j'(j(A))))$. Thus $(J' \circledcirc J)(A) \in$
$\in \text{I}(Q(A),S((j'oj)(A)))$. This proves that $(J' \circledcirc J) \in \text{Im}(Q,S)$. ♦

Lemma 9.2.3: If $J \in \text{Im}(R,S)$ then $J^i \in \text{Im}(S,R)$.
Proof: $J^i(A)=J(j^{-1}(A))^{-1}$. For all $X \in \mathbb{D}_R$, $J(X) \in \text{I}(R(X),S(j(X)))$. Thus,
$J(j^{-1}(A)) \in \text{I}(R(j^{-1}(A)),S(j(j^{-1}(A))))$ which implies $J(j^{-1}(A)) \in \text{I}(R(j^{-1}(A)),S(A))$ and
from this it follows that $J^i(A) \in \text{I}(S(A),R(j^{-1}(A)))$. Since the first half of J^i is j^{-1}, it
follows that $J^i \in \text{Im}(S,R)$. ♦

Theorem 9.2.4: $\langle \text{Im}(R), \circledcirc, {}^i, \Upsilon \rangle$ is a group.
Proof: If $J,J' \in \text{Im}(R)$ then, according to 9.2.2, $(J' \circledcirc J) \in \text{Im}(R)$. If $J \in \text{Im}(R)$ then
$J^i \in \text{Im}(R)$ according to 9.2.3. Thus, $\text{Im}(R)$ is a subgroup of $\text{Bim}(\mathbb{D}_R)$. ♦

In section 9.1, the notion of a conservative mapping was introduced. A conservative mapping is of the form $\langle \iota_M, J \rangle$, and therefore for all $A \in M$, $J(A):A \rightarrow A$. A bijective mapping J on M into M is thus conservative iff $J(A) \in \text{Bi}(A)$ for all $A \in M$. We denote the set of conservative bijective mappings on M by $\text{CBim}(M)$. A conservative isomorphic mapping of R to S is a conservative bijective mapping J on $\mathbb{D}_R$ into $\mathbb{D}_S$ such that for all $A \in \mathbb{D}_R$, $J(A) \in \text{I}(R(A),S(A))$, and we denote the set of all such mappings by $\text{CIm}(R,S)$. The set of all conservative automorphic mappings of R is denoted $\text{CIm}(R)$. Υ is of course a conservative mapping.

Theorem 9.2.5: $\langle \text{CBim}(M), \circledcirc, {}^i, \Upsilon \rangle$ and $\langle \text{CIm}(R), \circledcirc, {}^i, \Upsilon \rangle$ are groups.
Proof: The composition of two conservative mappings is a conservative mapping. And the inverse of a conservative mapping is a conservative mapping. The theorem follows easily from these two facts. ♦

In section 9.4 we shall study another special kind of isomorphism mappings, global isomorphic mappings.

9.3 Automorphic mapping invariance

The notion of automorphism invariance, i.e. invariance under automorphisms, is an important notion in the context of invariance of relations. We have dealt with it in chapter 1 and 2. We shall now generalize it to the context of relationals, and arrive at the notion of *automorphic mapping invariance*. Let R and S be systems of relationals with the same range of definition D. We say that S is *automorphic-mapping-invariant* relative to R if $Im(R) \subseteq Im(S)$.

Theorem 9.3.1: Suppose R and S are relational systems with the same range of definition. Then S is automorphic-mapping-invariant relative to R iff S is subordinate to R, i.e. $Im(R) \subseteq Im(S)$ iff $I_R \subseteq I_S$.

Proof: Let the range of definition of R and S be D.

(I) Suppose $Im(R) \subseteq Im(S)$ and $\varphi \in I_R(A,B)$. Define the mapping $<j,J>$ in the following way:

$$j(X) = \begin{cases} B & \text{if } X=A \\ A & \text{if } X=B \\ X & \text{if } X \neq A,B \end{cases}$$

$$J(X) = \begin{cases} \varphi & \text{if } X=A \\ \varphi^{-1} & \text{if } X=B \\ \iota_X & \text{if } X \neq A,B \end{cases}$$

Obviously, $<j,J> \in Bim(D)$. Since $J(A)=\varphi$ and $j(A)=B$, it follows that $J(A) \in I_R(A,j(A))$. Since $J(B)=\varphi^{-1}$ and $j(B)=A$, it follows that $J(B) \in I_R(B,j(B))$. For all $X \in D$ such that $X \neq A,B$ it holds that $J(X) \in I_R(X,j(X))$ since $J(X)=\iota_X$ and $\iota_X(X)=X$. Thus, $J \in Im(R)$ and from the assumption follows $J \in Im(S)$. This implies that $J(A) \in I_S(A,j(A))$, and since $J(A)=\varphi$ and $j(A)=B$ we have $\varphi \in I_S(A,B)$. This shows that $I_R(A,B) \subseteq I_S(A,B)$.

(II) Suppose $I_R(A,B) \subseteq I_S(A,B)$ for all $A,B \in D$. Suppose further that there exists $J \in Im(R)$ such that $J \notin Im(S)$. Since $j \in Bi(D)$ there must exist $A \in D$ such that $J(A) \notin I_S(A,j(A))$. But $J(A) \in I_R(A,j(A))$, which implies, according to the assumption, that $J(A) \in I_S(A,j(A))$. We thus get a contradiction, so $Im(R) \subseteq Im(S)$. ♦

Theorem 9.3.1 establishes beyond any reasonable doubt that subordination is a kind of invariance condition.

Subordination is defined in terms of isomorphism preservation. Let us now introduce the notion of pointwise subordination which we define in terms of automorphism preservation. First a note on notation. We use $\mathbb{I}_R(A)$ to denote $\mathbb{I}_R(A,A)$, i.e. $\mathbb{I}(R(A),R(A))$. Let R and S be relational systems with the same range of definition $\mathbb{D}$. We say that S is *pointwise subordinate* to R—in symbols, Sp–$sub R$—if for all $A,B \in \mathbb{D}$,

$$\mathbb{I}_S(A) \supseteq \mathbb{I}_R(A).$$

It is shown in the next theorem that pointwise subordination is equivalent to invariance under conservative automorphic mappings.

Theorem 9.3.2: Suppose R and S are relational systems with the same range of definition $\mathbb{D}$. Then S is conservative automorphic-mapping-invariant relative to R iff S is pointwise subordinate to R, i.e. $\mathbb{C}\mathbb{Im}(R) \subseteq \mathbb{C}\mathbb{Im}(S)$ iff $\mathbb{I}_R(A) \subseteq \mathbb{I}_S(A)$ for all $A \in \mathbb{D}$.

Proof: (I) Suppose $\mathbb{C}\mathbb{Im}(R) \subseteq \mathbb{C}\mathbb{Im}(S)$ and $\varphi \in \mathbb{I}_R(A)$. Define the conservative mapping $\langle \iota, J \rangle$ in the following way:

$$J(X) = \begin{cases} \varphi & \text{if } X = A \\ \iota_X & \text{if } X \neq A \end{cases}$$

Since $J(A) = \varphi$ it follows that $J(A) \in \mathbb{I}_R(A)$. For all $X \in \mathbb{D}$ such that $X \neq A$ it holds that $J(X) \in \mathbb{I}_R(X)$. Thus, $J \in \mathbb{C}\mathbb{Im}(R)$ and from the assumption it follows that $J \in \mathbb{C}\mathbb{Im}(S)$. Therefore $J(A) \in \mathbb{I}_S(A)$, i.e. $\varphi \in \mathbb{I}_S(A,B)$. This shows that $\mathbb{I}_R(A) \subseteq \mathbb{I}_S(A)$.
(II) Suppose $\mathbb{I}_R(A) \subseteq \mathbb{I}_S(A)$ for all $A \in \mathbb{D}$. Suppose further that there exists $J \in \mathbb{C}\mathbb{Im}(R)$ such that $J \notin \mathbb{C}\mathbb{Im}(S)$. Thus, there must exist $A \in \mathbb{D}$ such that $J(A) \notin \mathbb{I}_S(A)$. But $J(A) \in \mathbb{I}_R(A)$, which implies, according to the assumption, that $J(A) \in \mathbb{I}_S(A)$. We thus get a contradiction, so $\mathbb{C}\mathbb{Im}(R) \subseteq \mathbb{C}\mathbb{Im}(S)$. ♦

9.4 Global isomorphic mappings and global subordination

In this section we study isomorphic mappings which are global and invariance under global automorphic mappings. By way of introduction we demonstrate that the set of all global bijective mappings on a set into itself constitutes a group.

Theorem 9.4.1: $\langle \mathbb{G}\mathbb{B}\mathrm{im}(M), \copyright, {}^i, \Upsilon \rangle$ is a group.

Proof: GBim(M) is a subset of Bim(M). Υ is obvious a global mapping. If J,J'∈ GBim(M) then, according to 9.1.12, (J'©J)∈ GBim(M). If J∈ GBim(M) then J^i∈ GBim(M) according to 9.1.10. Thus, GBim(M) is a subgroup of Bim(M). ◆

An isomorphic mapping on **R** to **S** which is global is called a global isomorphic mapping. The set of all global isomorphic mappings on **R** to **S** is denoted GIm(**R**,**S**). And the set of all global automorphic mappings on **R** is denoted by GIm(**R**). It is easy to see that

$$GIm(\mathbf{R},\mathbf{S}) = GBim(\mathbb{D}_\mathbf{R}, \mathbb{D}_\mathbf{S}) \cap Im(\mathbf{R},\mathbf{S}). \qquad (9.4.1)$$

The proof of the following lemma, which is omitted, is straightforward.

Lemma 9.4.2: Suppose J∈ GIm(**Q**,**R**) and J'∈ GIm(**R**,**S**). Then (J'©J)∈ GIm(**Q**,**S**) and J^i∈ GIm(**R**,**Q**).

9.4.1 implies that

$$GIm(\mathbf{R}) = GBim(\mathbb{D}_\mathbf{R}) \cap Im(\mathbf{R}).$$

Note that both GBim($\mathbb{D}_\mathbf{R}$) and Im(**R**) are subgroups of the group Bim($\mathbb{D}_\mathbf{R}$). From this the next theorem follows immediately, the intersection of two subgroups of a group being a subgroup of that group.

Theorem 9.4.3: $\langle$ GIm(**R**),©, ^{i},$\Upsilon\rangle$ is a group.

According to 9.1.11, a bijective mapping J on M into M' is global if there is a one-to-one function G on ∪M onto ∪M' such that for all A∈ M, G[A]=j(A) and G⌐A=J(A). It is often natural to represent the global bijective mapping J with the function G. Let as therefore say that a one-to-one function G on ∪M onto ∪M' is a global bijective mapping on M into M' if {G[A] | A∈ M}=M'.

Note that J is a global isomorphic mapping on **R** to **S** iff there is a one-to-one function G on ∪ $\mathbb{D}_\mathbf{R}$ onto ∪ $\mathbb{D}_\mathbf{S}$ such that for all A∈ M, G[A]=j(A) and G⌐A=J(A) and (G⌐A)∈ I(**R**(A),**S**(G[A])). Let us therefore say that a one-to-one function G on ∪ $\mathbb{D}_\mathbf{R}$ onto ∪ $\mathbb{D}_\mathbf{S}$ is a global isomorphic mapping of **R** to **S** if {G[A] | A∈ $\mathbb{D}_\mathbf{R}$}=$\mathbb{D}_\mathbf{S}$ and for all A∈ M, (G⌐A)∈ I(**R**(A),**S**(G[A])).

Subordination is invariance under automorphic mappings. We shall now study invariance under global automorphic mappings. Let **R** and **S** be systems of relationals with the same range of definition $\mathbb{D}$. We say that **S** is *globally subordinate* to **R** iff $\mathbb{GIm}(\mathbf{R}) \subseteq \mathbb{GIm}(\mathbf{S})$. That **S** is globally subordinate to **R** we write for simplicity as **S**g-sub**R**.

Theorem 9.4.4: If S*sub*R then S*g-sub*R.
Proof: Suppose S*sub*R. Then according to 9.3.1, $\mathbb{Im}(\mathbf{R}) \subseteq \mathbb{Im}(\mathbf{S})$. Suppose now that $J \in \mathbb{GIm}(\mathbf{R})$. Then $J \in \mathbb{Im}(\mathbf{S})$ and J global. Thus $J \in \mathbb{GIm}(\mathbf{S})$. ♦

In section 4.4 it was shown that the regionalization R* is not generally subordinate to R, although R* is in a sense constructed or defined from R. This may seem strange, but the following theorem makes the situation somewhat clearer.

Theorem 9.4.5: Suppose R is an n-ary local relational. Then the regionalization of R is globally subordinate to R, i.e. R**g-sub*R.
Proof: Suppose $J \in \mathbb{GIm}(R)$. Then we can represent J by J#, and $J\# \in \mathbb{Bi}(\cup \mathbb{D})$ where $\mathbb{D}$ is the range of definition of R. It holds that
$$R^*(A; a_1,...,a_n) \text{ iff } R(\{a_1,...,a_n\}; a_1,...,a_n) \text{ iff}$$
$$\text{iff } R(\{J\#(a_1),...,J\#(a_n)\}; J\#(a_1),...,J\#(a_n)) \text{ iff } R^*(J\#[A]; J\#(a_1),...,J\#(a_n)).$$
This shows that $(J\#\ulcorner A) \in \mathbb{I}(R^*(A), R^*(J\#[A]))$. Thus $J \in \mathbb{GIm}(R^*)$. ♦

g-sub expresses a condition of complete structural dependence but not of the same kind as *sub:* although sub expresses a domainwise dependence, this does not hold for *g-sub*. If S*sub*R then the structure of S(A) is determined by the structure of **R**(A). If S*g-sub*R, then in general it is not just the structure of **R**(A) that determines S(A), but rather the structure of all the extensions of **R** that determines each S(A).

g-sub expresses a kind of complete dependence. A plausible conjecture might therefore be that a generalized notion of uncorrelation—*g-uncorr*—can be defined which generates a dependence scale with *g-sub* at one extreme and *g-uncorr* at the other.

FINAL REMARKS

We begin this, the last section, with a quotation from W. E. Johnson (1922) pp. 219-220.

> The variations of the phenomenal character P depend *only* upon variations in the characters A, B, C, D,E (say).
>
> The conception of dependence, which the above formula introduces, requires more precise explanation. In the first place the formula must be understood to imply that the variations of A, B,C, D, E, upon which variations of P depend, are independent of one another. For if, for example, a variation of A entailed a variation of B, then B being a *determined* character should be omitted from amongst the *determining* characters. It is only by observing this principle that we can apply the essential rule for all experimentation—that *one* only of the determining characters should be varied at a time. Again it is essential that A, B, C, D, E, should be simplex characters: for the nature of the dependence of P upon them is such that, if only one of these mutually independent determining characters varies, the character P will vary; whereas, if more than one of them varied, P might remain constant. This consideration shows that if any character such as A was not simplex, but resolvable into unknown factors X and Y which varied independently of one another, then a variation in A might involve such a variation in both X and Y that the character P would remain unchanged.

This quotation serves to illustrate the lines along which philosophers have discussed a notion of dependence closely related to that of variation. Contributors include Jevons, Stebbing, Wisdom and Marc-Wogau, as well as the grand old man of the subject, John Stuart Mill. The discussion has waned during the last decades. Trains of thought closely allied to this discussion are to be found in the literature on measurement theory—for example in connection with conjoint measurement and decomposability—but with, as far as I know, few references to this older discussion. My thought has been that it is nevertheless possible to study more explicitly, within the general framework of modern measurement theory, the interplay between dependence and variation and thereby to reestablish a link with the older discussion.

Johnson says in the above quotation that if a variation of A entails a variation in B, then B is a determined character. This means, I think, that Johnson would accept that if a variation of A entails a variation in B, then B is determined by A. Now while this is not an uncommon idea in connection with dependence, it is neither in accordance with the idea of subordination nor with that of conformity. As was pointed out in section 4.8, if β is subordinate to α, then a variation of β implies a variation of α. It is not the case, given that β is subordinate to α, that a variation of α implies a variation of β. (In this informal discussion I express myself in terms of aspects and not in terms of systems of relationals.) The situation in the case of conformity is analogous. That it is sometimes maintained that if β is determined by α then a variation of α implies a variation of β, is due, I suppose, to a confusion between the following two ideas:

(1) the variation of α determines the variation of β

(2) a variation of α implies a variation of β.

It seems reasonable to take (1) to describe subordination or conformity, according as the variation in question concerns structures or objects. For suppose we vary what holds of the object a with respect to α so that a and b become equal in α. Given (1), it would seem to follow that a and b are equal in β. That is to say, β conforms to α. (2), on the other hand, means that if a and b are not equal in α then a and b are not equal in β either. Contraposed, this reads: If a and b are equal in β, then they are also equal in α, which implies that α conforms to β. Together (1) and (2) give, as far as I can see, equiformity and parity. That a variation of α implies a variation of β is therefore often understood to mean the special kind of determination that equiformity and parity amounts to.

Dependence between aspects in elementary physics seems in the general case to be of an "all or nothing" character. This was discussed in section 6.1. If $h_0 = f \circ \langle h_1, h_2 \rangle$ where h_i is a measure for α_i, then a familiar situation in elementary physics is that α_i is equality preserving relative to $(\alpha_j \alpha_k)$ whenever i,j and k are distinct, and that α_i and α_j are independently realizable when i and j are distinct. One may wonder why this should be so. I believe that "the essential rule for all experimentation", to use Johnson's expression, is of fundamental importance for understanding this problem. The rule says "that *one* only of the determining characters should be varied at a time", and something like it seems to underlie experimentation in elementary physics. Is the situation analogous in more advanced physics? The problem is complicated, I fear, and it is not possible for me to enter into it here. Let me just point out the possibility that in special branches of physics the dependence relations expressed by laws need not perhaps have this "all or nothing" character, so that instances of partial dependence may be found.

A theory of structural dependence may have several applications. One is in the study of the language of physics which I have just touched upon. Another which I found of special interest is in creating a framework in terms of which trade offs and aggregations of different aspects or factors can be considered. In most decision situations the decision agent is faced with the problem of handling a large body of information. It is, for example, often necessary to consider what holds of the decision alternatives with regard to several different aspects. Psychological studies seem to indicate that to do this in an intuitive manner is seldom sufficient. The human being seems to have severe shortcomings as an "information integrator". To assist decision agents in situations where many factors must be united to an over all judgement, a conceptual framework in terms of which the final judgement could be reached step by step might be of value. For such a conceptual framework a theory of structural dependence seems to me to be indispensable. In this context partial dependence is in all likelihood of central importance.

REFERENCES

Arrow, K.J. *Social choice and individual values.* New York: Wiley, 1951, 2nd ed. 1963.

Bell, E.T. *The development of mathematics.* New York: McGraw-Hill, 1945.

Bergholm, C. *Lärobok i fysik för realgymnasiet 1. Allmän fysik.* Uppsala: J.A. Lindblads förlag, 1963.

Berkeley, E.C. *Mathematics for the intelligent non-mathematician.* New York: Simon & Schuster, 1966.

Beth, W.E. On Padoa's method in theory of definition. *Indagationes Mathematicae,* 1953, **15**, 330-339.

Birkhoff, G. On groups of automorphisms. (Spanish). *Rev. Un. Math.* Argentina, 1946, **11**, 155-157.

Birkhoff, G. *Lattice theory.* Amer. Math. Soc. Coll. Publ., Vol.25, New York: 3rd ed. 1967.

Blumenthal, L.M. & Menger, K. *Studies in geometry.* San Francisco: Freeman, 1970.

Bonevac. D. Supervenience and ontology. *American Philosophical Quarterly,* 1988, **25**, 37-47.

Carnap, R. *Meaning and necessity.* Chicago: The University of Chicago Press, 1947.

Carnap, R. Notes on semantics. *Philosophia,* 1972, Vol. **2**, Nos. 1-2, 1-54.

Chisholm, J.S.R. & Morris, R.M. *Mathematical methods in physics.* Amsterdam: North-Holland, 1965, 2nd ed. 1966.

Cohn, P.M. *Universal algebra.* New York: Harper & Row, 1965

Danielsson, S. *Two papers on rationality and group preferences.* Filosofiska studier. Uppsala: Filosofiska Föreningen och Filosofiska institutionen vid Uppsala Universitet, no 21, 1974.

Einstein, A. *Relativity, the special and general theory. A popular exposition.* 1916. Revised ed., London: Methuen, 1954.

Fishburn, P.C. *Utility theory for decision making.* New York: Wiley, 1970.

Fishburn, P.C. *The theory of social choice.* Princeton, New Jersey: Princeton University Press, 1973

Frege, G. *Translations from the philosophical writings of Gottlob Frege.* Ed. by Peter Geach and Max Black. Oxford, 1952.

Gallin, D. *Intensional and higher- order modal logic.* Amsterdam: North-Holland, 1975.

Gericke, H. & Martens, H. Some Basic Concepts for a Theory of Structure. In Behnke, Bachmann, Fladt & Süss (eds): *Fundamentals of Mathematics*. Vol.1. Cambridge, Massachusetts and London: MIT Press, 1974.

Grätzer, G. *Universal algebra*. 2nd ed. New York: Springer-Verlag, 1979.

Gårding, L. *Encounter with mathematics*. New York: Springer-Verlag, 1977.

Gärdenfors, P. Positionalist voting functions. *Theory and Decision*, 1973, **4**, 1-24.

Hansson, B. Group preferences. *Econometrica*, 1969, **37**, 50-54.

Hu, S-T. *Elements of general topology*. San Francisco: Holden-Day, 1964.

Jevons, W.S. *The principles of science. A treatise on logic and scientific method*. London: Macmillan, 1874, 2nd ed. reprinted 1924.

Johnson, W.E. *Logic*.Part II. Cambridge: Cambridge University Press, 1922.

Kanger, S. *Mätning. En vetenskapsteoretisk essay*. Filosofiska studier. Stockholm, 1963.

Kanger, S. Measurement: an essay in philosophy of science. *Theoria*,1972, **38**, 1-44.

Keeney, R.L. & Raiffa, H. *Decisions with multiple objectives: preferencesand value tradeoffs*. New York: Wiley, 1976.

Keyser, C.J. Hibbert Journ.,1904-5, **3**, 313.

Klein, F. *Vergleichende Betrachtungen über neuere geometrische Forschungen*. Erlangen, 1872.

Klein, F. *Gesammelte mathematische Abhandlungen, Erster Band*, Berlin: Verlag von Julius Springer, 1921.

Krantz, D.H., Luce, R.D., Suppes, P. & Tversky, A. *Foundations of measure ment*.Vol I. New York: Academic Press, 1971.

Krantz, D.H., Luce, R.D., Suppes, P. & Tversky, A. *Foundations of measurement*.Vol II. Forthcoming.

Kunle, H. & Fladt, K. Erlanger Program and Higher Geometry. In Behnke, Bachmann, Fladt & Kunle (eds.): *Fundamentals of Mathematics,Volume II, Geometry*. Cambridge, Massachusetts: MIT Press, 1974.This is an English translation of the German original *Grundzüge der Mathematik*. Göttingen: Vandenhoeck & Ruprecht, 1967 and 1971.

Kuratowski, K. *Introduction to set theory and topology*. Second edition. Oxford: Pergamon Press, 1972

Luce, R.D. Semiorders and a theory of utility discrimination. *Econometrica*,1956, **24**,178-191.

Luce, R.D. & Raiffa, H. *Games and decision*. New York: Wiley, 1957.

Luce, R.D.& Suppes, P. Preference, Utility, and Subjective Probability. In Luce, Bush & Galanter (eds.):*Handbook of Mathematical Psycology. Vol 3*. New York: Wiley, 1965.

Marc-Wogau, K. *Modern logik*. Stockholm, 1950.

Margenau, H. Some integrative principles of modern physics. In Margenau (ed.):

Integrative principles of modern thought. 1972.

Menger, K. On variables in mathematics and in natural science. *Brit. J. Phil. Sci.,*1954, **5**, no18.

Menger, K. What are variables and constants? *Science,* 1956, **123**, 547-548.

Menger, K. Variables, constants, fluents. In H. Feigl & G. Maxwell (ed.): *Current issues in the philosophy of science.* Holt, Rinehart and Winston. New York, 1961.

Mill, J.S. *A system of logic.* Book III. London, 1843.

Montague, R. *Formal Philosophy. Selected papers of Richard Montague.*Ed. Richmond Thomason. New Haven: Yale University Press, 1974.

Pfanzagl, J. *Theory of measurement.* Vienna, 1968, 2nd ed. 1971.

Rabinowicz, W. Act-utilitarian prisoner's dilemmas. *Theoria*, 1989, **15**, 1-44.

Roberts, F.S. *Measurement theory with applications to decisionmaking, utility, and the social sciences.* Reading, Massachusetts: Addison-Wesley, 1979.

Scott, D. & Suppes, P. Foundational aspects of theories of measurement. *J. Symbolic Logic,* 1958, **23**, 113-128.

Sen, A.K. *Collective choice and social welfare.* San Francisco: Holden-Day, 1970.

Shoenfield, J. *Mathematical logic.* New York: Addison-Wesley, 1967.

Stebbing, L.S. *A modern introduction to logic.* London: Methuen, 1930, 2nd ed. 1933.

Stevens, S.S. Mathematics, measurement, and psychophysics. In Stevens (ed.): *Handbook of experimental psychology.* New York: Wiley, 1951.

Stevens, S.S. Measurement, statistics, and the schemapiric view. *Science*, 1968, **161**, 849-856.

Stoll, R.R. *Set theory and logic.* San Francisco: Freeman, 1963.

Suppes, P. *Introduction to logic.* Princeton, New Jersey: van Nostrand, 1957.

Suppes, P. *Set-theoretical structures in science.* Mimeographed. Stanford: Stanford University, Institute for Mathematical Studies in the Social Sciences, 1967.

Suppes, P. & Zinnes, J.L. Basic measurement theory. In R.D. Luce, R.R. Bush, & E. Galanter (eds.): *Handbook of mathematical psycology.* Vol 1. New York: Wiley, 1963.

Todhunter, I. *A treatise on the differential calculus.* London, 1852.

Weyl, H. *Philosophy of mathematics and natural science.* Princeton: Princeton University Press, 1949.

Weyl, H. *Space, time, matter.* 1922. Reprinted in Dover Publications, New York, 1952.

Wisdom, J.O. Criteria for causal determination and functional relationship. *Mind*, 1945, **54**, ser 2.

INDEX